Th. W. Bakker, P. D. Jungerius & J. A. Klijn (Editors)

DUNES OF THE EUROPEAN COASTS

Geomorphology – Hydrology – Soils

CATENA SUPPLEMENT 18

CATENA – A Cooperating Journal of the International Society of Soil Science

Cover photo by Erik Wanders: Dunes on the Hebrides

CIP-Titelaufnahme der Deutschen Bibliothek

Dunes of the European coasts : geomorphology - hydrology - soils / Th. W. M. Bakker . . . (ed.). - Cremlingen-Destedt : CATENA-Verl., 1990
(Catena : Supplement; 18)
ISBN 3-923381-23-9
NE: Bakker, Th. W. M. [Hrsg.]; Catena / Supplement

©Copyright 1990 by CATENA VERLAG, D-3302 CREMLINGEN-Destedt, W. GERMANY

All rights are reserved. No part of this publication may be reproduced, stored in a retrieval system or transmitted in any form or by any means, electronic, mechanical, photocopying, recording or otherwise, without prior permission of the publisher.

This publication has been registered with the Copyright Clearance Center, Inc. Consent is given for copying of articles for personal or internal use, for the specific clients. This consent is given on the condition that the copier pay through the Center the per-copy fee for copying beyond that permitted by Sections 107 or 108 of the U.S. Copyright Law. The per-copy fee is stated in the code-line at the bottom of the first page of each article. The appropriate fee, together with a copy of the first page of the article, should be forwarded to the Copyright Clearance Center, Inc., 27 Congress Street, Salem, MA 01970, U.S.A. If no code-line appears, broad consent to copy has not been given and permission to copy must be obtained directly from the publisher. This consent does not extend to other kinds of copying, such as for general distribution, resale, advertising and promotion purposes, or for creating new collective works. Special written permission must be obtained from the publisher for such copying.

Submission of an article for publication implies the transfer of the copyright from the author(s) to the publisher.

ISSN 0722-0723 / ISBN 3-923381-23-9

CONTENTS

Editorial

European coastal dunes from the bird's eye

J.A. Klijn
Dune forming factors in a geographical context   1
E.C.F. Bird
Classification of European dune coasts   15

Regional studies in geomorphology

R.K. Borówka
Coastal dunes in Poland   25
R.W.G. Carter
The geomorphology of coastal dunes in Ireland   31
M. Robertson-Rintoul & W. Ritchie
The geomorphology of coastal sand dunes in Scotland: a review   41
H. Tsoar
Trends in the development of sand dunes along the southeastern Mediterranean coast   51

Dating of dune systems

Ch. Christiansen, K. Dalsgaard, J.T. Møller & D. Bowman
Coastal dunes in Denmark; chronology in relation to sea level   61
P. Wilson
Coastal dune chronology in the north of Ireland   71
M.J. Tooley
The chronology of coastal dune development in the United Kingdom   81
J.A. Klijn
The younger dunes in The Netherlands; chronology and causation   89
C. Bressolier, J.-M. Froidefond & Y.-F. Thomas
Chronology of coastal dunes in the south-west of France   101

Geohydrology

T.W.M. Bakker
The geohydrology of coastal dunes   109
P.J. Stuyfzand
Hydrochemical facies analysis of coastal dunes and adjacent low lands: The Netherlands as an example   121
T.W.M. Bakker & P.R. Nienhuis
Geohydrology of Les Dunes de Mont Saint Frieux, Boulonnais, France   133
M.R. Llamas
Geohydrology of the eolian sands of the Doñana National Park (Spain)   145

Aspects of dune soils

P.D. Jungerius
The characteristics of dune soils   155
H.J. Mücher
Micromorphology of dune sands and soils   163
L.W. Dekker & P.D. Jungerius
Water repellency in the dunes with special reference to the Netherlands   173

Erosion and stabilization

P.D. Jungerius & L.W. Dekker
Water erosion in the dunes   185
J.L.A. Pluis & B. De Winder
Natural stabilization   195

Future developments

F. van der Meulen
European dunes: consequences of climate change and sealevel rise   209

Epilogue
Appendix

Editorial

With the integration of the European Community there is a growing awareness among scientists that landscapes which the participating countries have in common should be studied within a European framework. One of these landscapes is formed by the coastal dunes found from northernmost Norway to the extreme south of Spain. Together with other coastal areas their economic and ecological importance is great. Moreover, man-induced or natural threats, such as large-scale construction, reclamation or sea level rise, ask for continual measures based on scientific insight.

Awareness that the coastal zone merits communal concern has led to the recent foundation of two organizations: the European Union for Dune Conservation and Coastal Management (EUDC) and Eurocoast. Both are active in pooling knowledge of the coastal zone, organizing meetings, and promoting other related matters. This Supplement of CATENA should be seen as another effort to stimulate interest in one of the most striking aspects of the coastal zone: the dune landscape.

The diversity of dune landscapes is astonishing. There is probably no other type of landscape in the world which in its two main elements, land form and land cover, shows so much variation over such short distances. The spectacular variety of form elements is above all a consequence of the combination of wind and sand, with the former representing a process capable of acting independent of gravity, and the latter constituting a substrate which deforms easily. The variety of land cover is related to plant-growth conditions which change drastically over short distances depending on differences in microclimate, groundwater regime, soil type and distance from the sea. Added to this there is the diversity caused by humans who intervene in a number of ways in their efforts to adapt the landscape to their needs. These needs spring from the many functions the dunes can have for society: recreation, building, nature conservation, protection against the sea, agriculture, water extraction, etc.

Thus far, coastal dune landscapes have received little attention from the scientists who usually fill the contents of CATENA with contributions concerning "soil science, hydrology and geomorphology, focusing on geoecology and landscape evolution". It is mostly from botanists and vegetation ecologists that we have gained some understanding of dune ecosystems. They were among the first who realized the importance of environmental stress to the development of dune-building plants, thus drawing our attention to abiotic processes active in the dune ecosystem.

The fact that few geomorphologists and pedologists have shown interest in the coastal dunes is amazing considering that dune landscapes form one of the few remaining natural regions not being located in remote and inaccessible corners of the world. Moreover, all sorts of geomorphological processes typical of widely different climates are active here at a truly exceptional speed. In this respect coastal dunes can be regarded as a genuine 'field laboratory'. The study of soil formation in dunes is much neglected even though pedological processes here are far more natural than are those found in agricultural regions with their man-made surface horizons. Those active in the dunes have noticed to their satisfaction that their studies are highly appreciated by the managers of dune terrains, owing to the relevance that their findings have to nature conservation and other dune functions.

Coastal dunes have scientific value to soil scientists, hydrologists and geomorphologists in at least two respects. First of all, there is a great need for a better understanding of the abiotic part of the ecosystems as a complement to the ecological research of biologists. Such studies should be carried out at three levels: those of individual plants (autecology), the entire vegetation complex (vegetation ecology), and the landscape (landscape ecology). Each of these levels requires an appropriate scale of approach. It is only by studying the biotic-abiotic relationships at these three levels that the functioning of the dune ecosystem will eventually be understood.

Secondly, coastal dunes deserve scientific attention from a disciplinary point of view. Few environments are more dynamic in terms of geomorphological, pedological and hydrological processes. For the geomorphologists there are processes not only of wind action but also of water erosion to be studied. Those interested in the genesis of landscapes, find opportunities to match their conclusions with the information provided by other disciplines like palynology, archaeology and historical geography. Pedologists can study processes like the formation of organic soil profiles and the beginning of podzolization. Hydrologists encounter important problems which are incompatible with accepted hydrological models, e.g. water repellency and other small-scale matters such as water quality as affected by micro-geomorphological patterns. All of them find a wide variety of conditions at different spatial or time levels.

For this CATENA SUPPLEMENT we have invited scientists from various countries to discuss briefly the 'state of the art' of their specialization, and to provide information on the kind of research they are presently involved in. In the few pages allotted them this was no mean task. The enthousiastic response clearly demonstrates that dune research has a bright future.

Theo Bakker
Peter Jungerius
Jan Klijn

The production of this Supplement has been made possible with the financial support of the Department of Nature Conservation, Environmental Protection and Wildlife Management, Ministry of Agriculture, Nature Management and Fisheries, The Netherlands.

DUNE FORMING FACTORS IN A GEOGRAPHICAL CONTEXT

J.A. **Klijn**, Wageningen

Summary

The distribution of coastal sand dunes in Europe logically reflects the role of four dune forming factors: sand, wind, sea and plant growth. On a macroscale, tectonic history and the sedimentary history of coastal areas and the shelf zone help to explain the availability of sand for coastal dune building. On a more regional scale, large and mainly fossil river systems act as sand sources. Beach development favouring dune forming processes is promoted in storm wave environments, where tidal ranges are large and wave energy is high. Dune building occurs when onshore winds are strong enough (over 4 Beaufort) and frequent. The best conditions are found along the Southeastern North Sea basin and the west facing Atlantic coasts of France, Great Britain and Portugal. The role of plant growth as a dune forming factor deserves further study on a European scale.

1 Introduction

Dunes can originate where wind energy is sufficiently strong to transport unconsolidated sediments of weathering residues, especially sand. Desert dunes can form where arid conditions prevail and consequently plant cover is sparse or even absent. Here, two factors control the dune forming processes:

i) the availability of sand and

ii) wind force and direction.

Although desert dunes are found near many coasts, such as in Namibia, Australia, North Africa, Peru, Mexico, California (DAVIES 1972, GOLDSMITH 1978), in the proper sense coastal dunes have a different origin. Four dune forming factors are chiefly involved:

i) the sand,

ii) the sea

iii) the wind and

iv) plant growth (GRIPP 1968, KLIJN 1981).

Prominent features of these dunes are: firstly the presence of beach deposits constituting a distinct source of sand, secondly the role of the sea in the transporting of sand towards the beaches or in the destruction of existing dunes, and thirdly, the essential role of plant growth in all stages of the dune forming processes. Consequently some authors introduced terms such as organogenic or phytogenic dune forming processes (VAN

ISSN 0722-0723
ISBN 3-923381-23-9
©1990 by CATENA VERLAG,
D-3302 Cremlingen-Destedt, W. Germany
3-923381-23-9/90/5011851/US$ 2.00 + 0.25

DIEREN 1934, SMITH 1935). The four dune forming factors already mentioned will be elaborated upon in the next section and their relative influence in relation to actual dune distribution will be evaluated in section 3.

2 Influence of geological, hydrographic, climatological and biological factors: a geographical approach

The presence and development of coastal dunes are strongly related to geological, climatological and hydrographic phenomena. The biological factor seems to be dependent on climatological and pedological factors. In this paragraph some relevant parameters are dealt with in their geographical context.

2.1 Sand

Apart from the availability of sand, its textural and chemical properties are considered to be important for dune formation and morphology. Availability and properties can vastly differ due to the tectonic and sedimentary history of the sea bed and the adjacent coastal area. The main factors governing the availability of sand are:

- width and depth of the shelf zone,
- sedimentary conditions of the sea bed,
- presence of sandy sediments in adjacent land areas,
- presence of major rivers carrying sand towards the coastal zone,
- abrasion of cliffs.

Fig. 1: *Distribution of continental shelves, indicated by the 150 m depth contour (map fragment from POSTMA & ZIJLSTRA 1988).*

In fig. 1 the shelf zone is shown for the European coasts indicated by the 150 m depth line. The coastline of NW-Europe, including the British Isles and the Baltic countries, is surrounded by a wide shelf. In contrast, the Iberian peninsula as well as a greater part of the Mediterranean coasts are bordered by a rather narrow shelf, with the exception of the Adriatic coasts. This large-scale distribution relates to the phenomena of plate tectonics, such as collision or trailing edges (fig. 2; INMAN & NORDSTROM 1971, EISMA 1988). The availability of sand in the sea bed or in adjacent land areas near the coast bears a striking correlation to erosional or sedimentary conditions during the early Holocene, Pleistocene and even Tertiary periods. During glacial periods in most northern and western European coastal

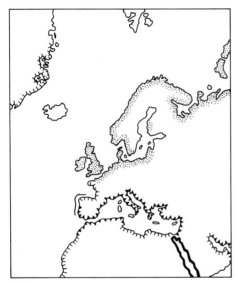

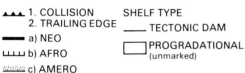

Fig. 2: *Tectonic classification of coasts and shelves (map fragment from DAVIES 1972 after data from INMAN & NORDSTROM 1971).*

areas glaical, fluvioglacial or eolian processes determined the sedimentary conditions in what nowadays is the shelf-zone (RANWELL 1972, DAVIES 1972, BIRD, this volume). This is emphasized by the fact that during the Weichselien, approx. 18000 years ago, the sea level was ca. 120 metres lower than it is now. It is known that in the case of the Saalien the level was about the same (120 to 130 m below the present level) (JELGERSMA 1979). In fig. 3 the conspicuous features are the presence of recently deglaciated and ice-scoured mountaineous areas in Scandinavia and Scotland. In these regions the earth's crust is subject to uplift due to postglacial debur-dening and southwardly there is a vast area where glacial and periglacial fluvial and eolian deposits abound. This zone includes most of the Baltic region and coastal areas of Northwest Europe as far as northern France. A gradually rising sea level due to isostatically determined subsidence of large area/sections of this zone, is typical (PIRAZZOLI 1985). These large scale patterns shown in fig. 3 are essentially zonal in character (BÜDEL 1977). Fig. 3 also shows the extent of glaciated areas in the Saalien, the Weichselien and the distribution of glacial sediments and outwashes or periglacial eolian sediments as well.

The same map shows the presence of major river systems, often accompanied by large basins filled with fluvial sediments, such as pebbles, sand or finer material. The large river systems are, for example, the Gironde and Rhone in southern France, the Po-delta in Italy and the Ebro, Taag and Guadalquivir in the Iberian peninsula. These systems often date back to at as early as the Tertiary. At present their importance as sand carrying agents is relatively small.

Logically, the abrasion of cliffs is most important in the case of unconsolidated sediments of glacial, periglacial or fluvial origin as described above. Besides, some coasts consist of easily erodible rock. These data are not included in fig. 3 (see also BIRD, this volume).

2.2 Sea

The influence of the sea as a dune forming factor is threefold:

i) it controls the sand budget,

ii) it promotes eolian processes on beaches by keeping the surface bare and

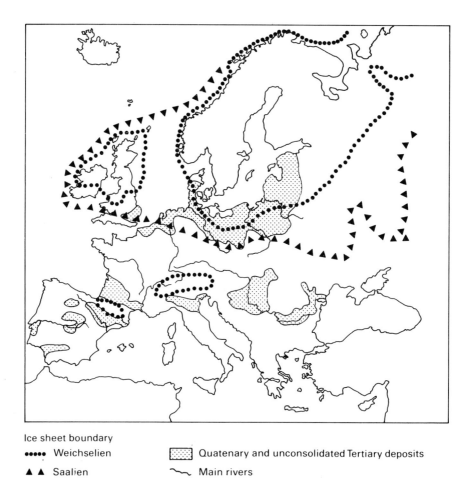

Fig. 3: *Major Quaternary and Tertiary sedimentary environments (free after FAO-UNESCO 1981 and EMBLETON 1984).*

iii) it affects the hinterland by sea-spray.

The most conspicuous feature is the influence the sea has on the sand budget along the coast. Waves and currents carry sand towards the coast, thus causing coastal accretion and subsequent dune building, yet in other regions or periods coastal erosion occurs. The effects can be catastrophic, particularly during storm surges. Whatever the cause, wave-cut cliffs in dunes are very susceptible to wind erosion and all patterns of secondary dune forming can be triggered by coastal erosion (VAN DIEREN 1934, KLIJN 1981, KLIJN, this volume).

The influence of the sea can be, furthermore, indirect due to the combined effects of tides, wave action and saline conditions of the beach environment. The combination of mechanical and chemical stress, reinforced by a severe microclimate causes a lack of vegetation. The resulting bare surface promotes eo-

Dune Forming Factors

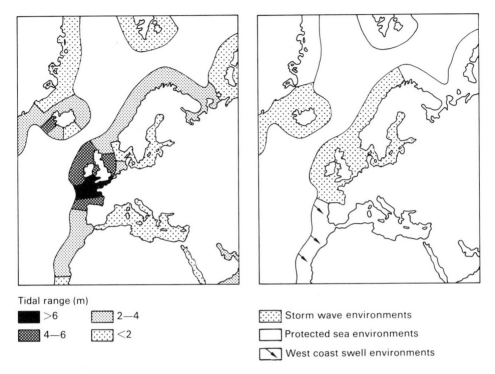

Tidal range (m)
- ■ >6
- ▓ 4—6
- ▒ 2—4
- ⋯ <2

▒ Storm wave environments
□ Protected sea environments
◣ West coast swell environments

Fig. 4: *Distribution of tidal range in metres at spring tide (map fragment from DAVIES 1972).*

Fig. 5: *Storm wave environments (map fragment from DAVIES 1972).*

lian processes. Yet another indirect influence of the sea is the effect of sea-spray on coastal dune systems, stimulated by wave action and onshore winds. The salt load can affect vegetation and dune morphology.

Consequently, crucial marine factors for the beach-dune systems are tidal range, tidal currents, the occurrence and intensity of storm floods and the energy and incidenc of wind generated waves. Tidal ranges as shown in fig. 4 strongly vary along European coasts. Notably low ranges occur in the Baltic region and in the Mediterranean sea, the open Atlantic coasts of Norway and Portugal show a limited tidal range (2–4 m), while the coasts of northern France, Belgium and around the British Isles experience rather high ranges (4-6 m; over 6 m, even up to 12 m) (SALOMON & ALLEN 1983).

Storm surges, leading sometimes to a considerable rise in the sea level occur in the zone where cyclonic depressions from the Atlantic Ocean abound. The Norwegian coast, Denmark, Germany, The Netherlands, Belgium, Northern France and the British Isles frequently suffer storm surges, especially where hydrograhic conditions such as in the southern North Sea or in the Channel, help to enlarge the effect of storm floods alone.

Wave energy, depending on the length and height of the waves, correlate with wind force and duration, fetch and depth. Wave energy increases strongly with wind force. The effect on coasts, in this case caused by parallel or perpendicular sand transport, relates to coastal exposition to dominant wave direction. Most European coasts experience strong winds from westerly quarters during the winter. The Atlantic coasts of Norway, the British Isles and the French and northern Iberian coast, are greatly exposed to high energy waves.

According to DAVIES (1972), conditions for coastal dune development are best in the so-called storm wave environments (fig. 5). Strong onshore winds, rather high wave energy and the occurrence of storm surges combine to create favourable conditions for sand supply and availability of sand for eolian processes on beaches.

2.3 Wind

Sand grains can be set in motion above a critical wind speed (4 m/s equals approx. 3 to 4 Beaufort windforce) (BAGNOLD 1954). Most of the particles move by saltation; less important is airborne transport or transport by rolling (SINDOWSKI 1956). Transport energy increases exponentially with wind speed. Wind force, wind direction and wind frequencies are important for primary dune building as well as for secondary dune forming (dune deformation, dune displacement). Wind forces over 4 Beaufort are capable of transporting sand (BAGNOLD 1954, ADRIANI & TERWINDT 1974). Therefore is it relevant to illustrate the distribution of the frequency of onshore winds with forces over 4 Beaufort in January and July (fig. 6) after DAVIES 1972). Due to transport energy considerably increasing with wind forces over 8 Beaufort, fig. 7 shows analogous data for onshore winds with a force of 8 Beaufort or more. A comprehensive view explains that the Atlantic region of Northwestern Europe in particular, experiences frequent strong onshore winds due to the abundance of cyclonic depressions in the westerlies north of the 45 degree latitude. The Mediterranean area, especially the most eastern as well as the most westerly parts are relatively quiet areas.

2.4 Plant growth

Plant growth plays a vital role from the very beginning of dune building on beaches or beach plains and throughout all the subsequent stages of secondary dune forming processes. Vertical dune growth in embryonal dunes on beach plains or in foredune ridges seems to be related to the tolerance of overblown pioneer species such as *Agropyron, Elymus* and *Ammophila*.

An annual rate up to 0.6 meter is thought to be the maximum for the last-mentioned species (RANWELL 1972). Apart from the very presence of these vital species (COOPER 1958), which in fact is a biogeographical factor, the extent to which environmental stress, e.g. cold or drought or the nutrient status, could limit vertical growth of these pioneer vegetations and the accompanying dunes is of significance. When existing plant cover cannot recover after being damaged, therefore enabling secondary dune forming processes, the same climatic and pedological constraints play a role. Deterioration of plant cover of course is often triggered by animal influences or abuse by man (KLIJN 1981).

Although biogeographical districts have been discerned based on floristic and synecological differences (GÉHU 1985, HEWETT 1989), in the literature

Dune Forming Factors

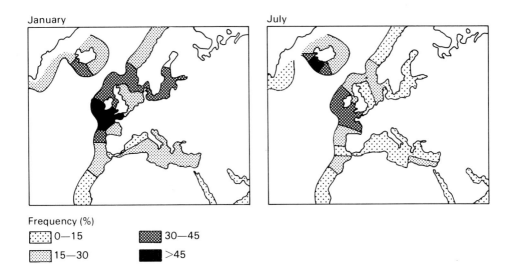

Fig. 6: *Onshore winds (in average % frequency) stronger than 4 Beaufort in January (a) and July (b). Map fragment from DAVIES (1972).*

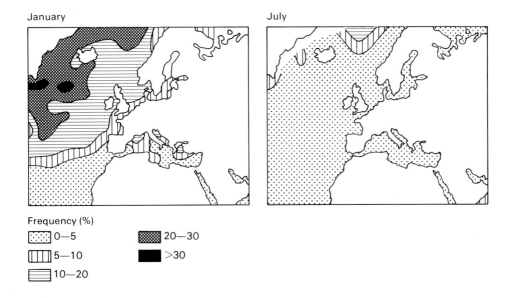

Fig. 7: *Frequency of occurrence (in %) of winds stronger than 8 Beaufort, in January (a) and July (b). Map fragments from DAVIES (1972).*

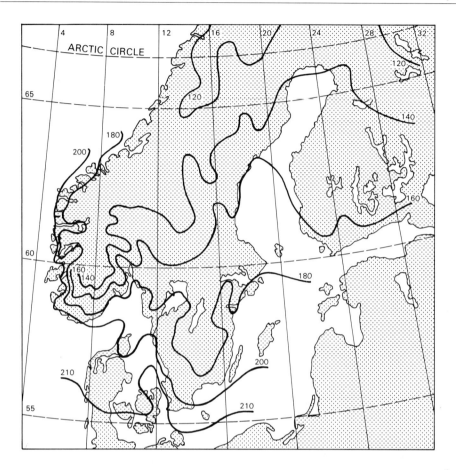

Fig. 8: *Length of vegetation period (daily mean temperature $\geq 6°C$) (after WALLÉN 1970)*.

relatively little attention has been paid to possible climatic or pedological restaints leading to marked differences in pioneer plant growth or vitality resulting in differences in dune morphology. Hypothetically the following climatological data are relevant: the length of the growing season, determined by cold stress (fig. 8), and drought stress determined by annual rainfall and evapotranspiration (fig. 9). Certain areas such as the south-eastern part of Spain are almost arid, with precipitation less than 300 mm and high evapotranspiration rates (900 mm). In addition, there are several regions with a rather long severe summer drought (fig. 10). It could be interesting to conduct correlative studies on the possible relationships of climatic restraints and the annual growth and density of the main pioneer communities on beaches and in foredunes along European coasts.

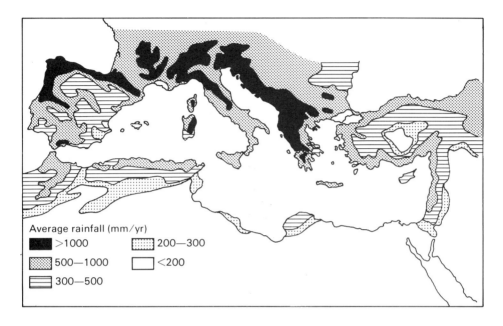

Fig. 9: *Average rainfall in the Mediterranean region (from PISSARIDES 1988; after data from the Times Atlas of the World).*

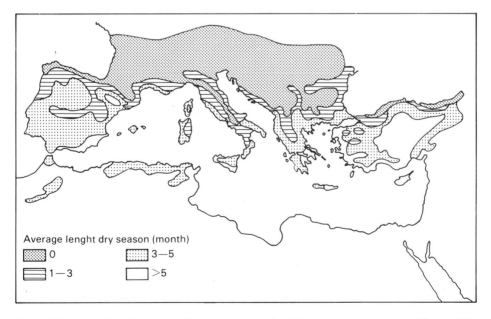

Fig. 10: *Length of average dry season in the Mediterranean region (from PISSARIDES 1988).*

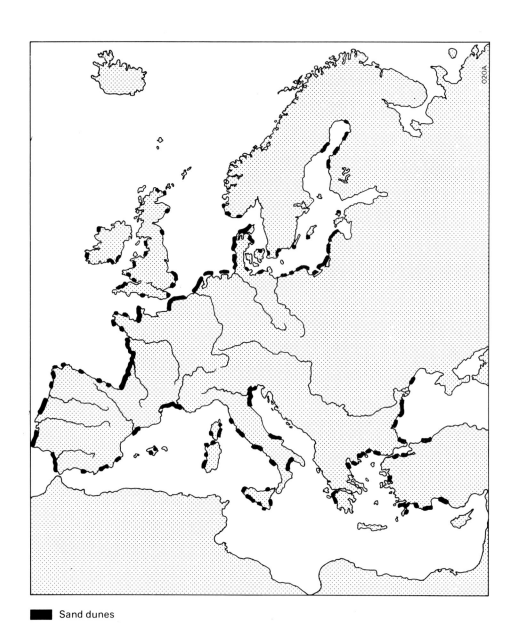

Fig. 11: *General distribution of dunes in western Europe (from GÉHU 1985).*

3 Distribution of European coastal dunes related to dune forming factors

Although national and international inventories have been carried out, such as in Denmark (KUHLMAN 1969), Scotland (MATHER & RITCHIE 1977), England and Wales (DOODEY 1989), The Netherlands (BAKKER et al. 1979), the North Sea coast from Calais (Northern France) to Skagen in Denmark (DEPUYDT 1972), Spain (BONNET FERNANDEZ TRUJILLO 1989), Turkey (USLU 1989) an overall survey of the occurrence and extent of coastal dunes along European coasts has yet to be made. The only smallscale map is provided by GÉHU (1985) as shown in fig. 11. A more specific and detailed inventory is in preparation at the moment under the supervision of the EUCD (European Union for Dune Conservation and Coastal Management). A qualitative comparison of GÉHU's data confirms the importance of the fourfactor approach discussed earlier and points to some interesting features.

Generally speaking, an almost uninterrupted zone of often extensive coastal dunes borders the lowlands of Northern France, Belgium, The Netherlands, N-Germany and the west coast of Denmark. Clearly in this zone Pleistocene and Holocene conditions guarantee abundant sandy deposits in both the North Sea Basin and subaerial regions. The shelfzone is wide and shallow, and the area is subject to frequent strong onshore winds and stormfloods. Slow subsidence seems to have had no negative effect. Maybe it ought to be regarded as a positive influence since slow subsidence over long geological periods could contribute to sedimentation of sand, whereas the effect on relative short term processes is subordinate to other influences such as sea level oscillations caused by climatic changes, changes in storm frequency or changes in management. This could also be said of parts of the coast of Normandy and Gascogne at the French Atlantic coast. In comparison with the mainland coast, the coastline of the British Isles is much more indented, and the dune areas are interrupted. Dune areas mainly occur in bays, river mouths entrances or other sheltered areas (TOOLEY, this volume). Coastal dunes along the coast of the Baltic countries and along the Bothnian Gulf show a preference for north or west facing coasts. GÉHU considers these dunes to be rather narrow (see for more specific information e.g. MISZALSKI 1973 or PIOTROWSKA 1989). Turning to the Iberian peninsula, the scarcity of coastal dunes on the northern coast could be related to a lack of sediment in the shelfzone and the absence of major fluvial sources. The availability of material, together with a more favourable exposure to westerly waves and winds may account for the valuable coastal dunes along parts of the Portuguese coasts. According to MARTINS (1989) 60% of the Portuguese coast is fringed by dunes.

The presence of an important fluvial source may explain the wealth of the coastal dunes near the Quadalquivir (Cota Doñana, SW Spain). Other dune areas along the Mediterranean coasts of Spain are more scattered, smaller and mostly explained by the relative sheltered situation of the coast or the presence of fluvial sediment sources. A more continuous strip of dunes is found near the Rhone Delta (Gulf of Lyon). Sedimentary conditions as well as a somewhat dynamic combination of tidal and climatic conditions here, could favour dune forming. According to GÉHU (1985) rather important, although narrow, dune areas are found on Corsica, Sardinia and Sicily. The mainland of Italy harbours rather small dune areas, a notable concentration being present in an explainable situation near the mouth of the river Po.

The coastline of Yugoslavia, Greece and Turkey are characterized by dune areas of lesser importance situated mainly in sheltered positions. Although USLU (1989) mentions a total area of 20,000 ha coastal dunes along the Turkish Mediterranean coast, this figure has to be considered in relation to the length of this coastline (1600 km) in order to make a meaningful comparison with other coasts in Europe, for instance The Netherlands, where ca. 40,000 ha coastal dunes are found along ca. 350 km coastline (BAKKER et al. 1979).

For the occurrence, extent and further explanation of dunes in Israel and the Negev, readers are referred to the contribution of TSOAR (this volume).

Acknowledgement

The author thanks Theo Bakker and Pim Jungerius for useful comments, B. ten Cate and Mrs. De Sylva for their textual improvements and G. van Dorland for preparing the drawings.

References

ADRIANI, M.J. & TERWINDT, J.H.J. (1974): Sand stabilization and dune building. Rijkswaterstaat publ. nr. 19, Staatsuitgeverij; Den Haag.

BAGNOLD, R.A. (1954): The physics of blown sand and desert dunes. London.

BAKKER, T.W.M., KLIJN, J.A. & VAN ZADELHOFF, F.J. (1979): Duinen en duinvalleien; een landschapsecologische studie van het Nederlandse duingebied. Pudoc, Wageningen, 201 pp.

BIRD, E.C.F. (1990): Classification of European dune coasts. This volume.

BONNET FERNANDEZ TRUJILLO, J. (1989): Dune management in Spain. In: F. van der Meulen, P.D. Jungerius & J.H. Visser (1989).

BÜDEL, J. (1977): Klima — Geomorphologie. Gebr. Borntraeger, Berlin-Stuttgart.

COOPER, W.S. (1958): Coastal sand dunes of Oregon and Washington. Geol. Soc. of America Memoir 72, 1–138.

DAVIES, J.L. (1972): Geographical variation in coastal development. Oliver & Boyd, Edinburgh.

DEPUYDT, F. (1972): De Belgische strand — en duinformaties in het kader van de geomorfologie der zuidoostelijke Noordzeekust. Kon. Ac. van Wetenschappen, Letteren en Schone Kunsten.

DOODEY, J.P. (1989): The conservation and development of coastal dunes in Great Britain. In: F. van der Meulen, P.D. Jungerius & J.H. Visser (1989).

EISMA, D. (1988): Geology of continental shelves. In: Postma, H. & Zijlstra, J.J. (Eds.), Ecosystems of the world; continental shelves. Elsevier, Amsterdam.

EMBLETON, C. (Ed.) (1984): Geomorphology of Europe. MacMillan Ref. Books, 465 pp.

FAO/UNESCO (1981): Soil map of the world 1:5 million. Vol. 5; Europe. Unesco, Paris.

GÉHU, J.M. (1985): European dune and shoreline vegetation. Council of Europe, Strassbourg. Brussel.

GRIPP, K. (1968): Zur jüngsten Erdgeschichte von Hörnum/Sylt und Amrum mit einer Übersicht über die Entstehung der Dünen in Nordfriesland. Die Küste Heft 16, 76–117.

GOLDSMITH, V. (1978): Coastal dunes. In: Davies, R.A. Jr. (Ed.), Coastal sedimentary environments. Springer Verlag.

HEWETT, D.G. (1989): Dunes and dune management on the Atlantic coast of Europe. In: F. van der Meulen, P.D. Jungerius and J.H. Visser (1989).

INMAN, D.L. & NORDSTROM, C.E. (1971): On the tectonic and morphologic classification of coasts. J. Geol. 79(1), 1–21.

JELGERSMA, S. (1979): Sea level changes in the North Sea basin. In: Oele, E., Schüttenhelm, R.T.E. & Wiggers, A.J. (Eds.), Quaternary history of the North Sea. Acta Univ. Symp. Univ. Upps., Uppsala, 223–248.

KLIJN, J.A. (1981): Nederlandse kustduinen; geomorfologie en bodems. Thesis, Wageningen.

KLIJN, J.A. (1990): The Younger Dunes in the Netherlands; chronology and causation. This volume.

KUHLMAN, H. (1969): Kyst, klit or marsk. In: Danmarks Natur, vol. 5.

MATHER, A.S. & RITCHIE, W. (1977): The beaches of the Highlands and Islands of Scotland. Countryside Commission of Scotland.

MARTINS, F. (1989): Morphology and management of dunes at Leira District (Portugal). In: F. van der Meulen, P.D. Jungerius & J.H. Visser (1989).

MISZALSKI, J. (1973): Present day aeolian processes on the Slovenian coastline. Warszawa IGPAN.

PIOTROWSKA, H. (1989): Natural and antropogenic changes in sand dunes and their vegetation on the Southern Baltic coast. In: F. van der Meulen, P.D. Jungerius & J.H. Visser (1989).

PIRAZZOLI, P.A. (1985): Sea level changes. Nature and Resources, Vol. XXI, no. 4, 2–9.

PISSARIDES, A. (1988): Development trends and environment programmes in Cyprus. In: Proc. Workshop and Conserv. and Development (Eds. Zomenis, S.L., Lüker, H. & Grievas, G.)

POSTMA, H. & ZIJLSTRA, J.J. (1988): Ecosystems of the world; continental shelves. Elsevier, Amsterdam.

RANWELL, D.S. (1972): Ecology of salt marshes and sand dunes. Chapman & Hall, London, 258 pp.

SALOMON, J.C. & ALLEN, G.P. (1983): Rôle sédimentologique de la marée dans les estuaria à fort marrage. Comp. Fr. Pétrol. Notes Mémoirs 18, 35–44.

SINDOWSKI, K.H. (1956): Korngrößen und Kornformen. Auslese beim Sandtransport durch Wind (nach Messungen auf Norderney). Geol. Jahrbuch, Band 71, Hannover.

SMITH, H.T.U. (1935): Classification of sand dunes. Congrès Géol. Int. Deserts actuels et anc. Sect. II, Fasc VII, 105 etc.

TOOLEY, M.J. (1990): The chronology of dune development in the United Kingdom. This volume.

TSOAR, J.H. (1990): Trends in the development of sand dunes along the southeastern Mediterranean coast. This volume.

USLU, T. (1989): Geographical information on Turkish coastal dunes. Report EUDC, Leiden.

VAN DER MEULEN, F., JUNGERIUS, P.D. VISSER, J.H. (Eds.) (1989): Perspectives in coastal dune management. SPB Ac. Publ. The Hague, NL, 333 p.

VAN DIEREN, J.W. (1934): Organogene Dünenbildung. Nijhoff, Den Haag.

WALLÉN, C.C. (1970): Climates of Northern and Western europe. In: World Survey of Climatology Vol. 5, Elsevier.

Address of author:
J.A. Klijn
The Winand Staring Centre
for Integrated Land, Soil and Water Research
P.O. Box 125
6700 AC Wageningen
The Netherlands

Jan de Ploey (Ed.)

RAINFALL SIMULATION, RUNOFF and SOIL EROSION

CATENA SUPPLEMENT 4, 1983

Price: DM 120,–/US $ 75,–

ISSN 0722–0723 ISBN 3-923381-03-4

This CATENA–Supplement may be an illustration of present-day efforts made by geomorphologists to promote soil erosion studies by refined methods and new conceptual approaches. On one side it is clear that we still need much more information about erosion systems which are characteristic for specific geographical areas and ecological units. With respect to this objective the reader will find in this volume an important contribution to the knowledge of active soil erosion, especially in typical sites in the Mediterranean belt, where soil degradation is very acute. On the other hand a set of papers is presented which enlighten the important role of laboratory research in the fundamental parametric investigation of processes, i.e. erosion by rain. This is in line with the progressing integration of field and laboratory studies, which is stimulated by more frequent feed-back operations. Finally we want to draw attention to the work of a restricted number of authors who are engaged in the difficult elaboration of pure theoretical models which may pollinate empirical research, by providing new concepts to be tested. Therefore, the fairly extensive publication of two papers by CULLING on soil creep mechanisms, whereby the basic force-resistance problem of erosion is discussed at the level of the individual particles.

All the other contributions are focused mainly on the processes of erosion by rain. The use of rainfall simulators is very common nowadays. But investigators are not always able to produce full fall velocity of waterdrops. EPEMA & RIEZEBOS give complementary information on the erosivity of simulators with restricted fall heights. MOEYERSONS discusses splash erosion under oblique rain, produced with his newly-built S.T.O.R.M-1 simulator. This important contribution may stimulate further investigations on the nearly unknown effects of oblique rain. BRYAN & DE PLOEY examined the comparability of erodibility measurements in two laboratories with different experimental set-ups. They obtained a similar gross ranking of Canadian and Belgian topsoils.

Both saturation overland flow and subsurface flow are important runoff sources under the rainforests of northeastern Queensland. Interesting, there, is the correlation between soil colour and hydraulic conductivity observed by BONELL, GILMOUR & CASSELLS. Runoff generation was also a main topic of IMESON's research in northern Morocco, stressing the mechanisms of surface crusting on clayish topsoils.

For southeastern Spain THORNES & GILMAN discuss the applicability of erosion models based on fairly simple equations of the "Musgrave-type". After Richter (Germany) and Vogt (France) it is TROPEANO who completes the image of erosion hazards in European vineyards. He shows that denudation is at the minimum in old vineyards, cultivated with manual tools only. Also in Italy VAN ASCH collected important data about splash erosion and rainwash on Calabrian soils. He points out a fundamental distinction between transport-limited and detachment-limited erosion rates on cultivated fields and fallow land. For representative first order catchment in Central–Java VAN DER LINDEN comments contrasting denudation rates derived from erosion plot data and river load measurements. Here too, on some slopes, detachment-limited erosion seems to occur.

The effects of oblique rain, time-dependent phenomena such as crusting and runoff generation, detachment-limited and transport-limited erosion including colluvial deposition are all aspects of single rainstorms and short rainy periods for which particular, predictive models have to be built. Moreover, it is argued that flume experiments may be an economic way to establish gross erodibility classifications. The present volume may give an impetus to further investigations and to the evaluation of the proposed conclusions and suggestions.

Jan de Ploey

G.F. EPEMA & H.Th. RIEZEBOS
FALL VELOCITY OF WATERDROPS AT DIFFERENT HEIGHTS AS A FACTOR INFLUENCING EROSIVITY OF SIMULATED RAIN

J. MOEYERSONS
MEASUREMENTS OF SPLASH–SALTATION FLUXES UNDER OBLIQUE RAIN

R.B. BRYAN & J. DE PLOEY
COMPARABILITY OF SOIL EROSION MEASUREMENTS WITH DIFFERENT LABORATORY RAINFALL SIMULATORS

M. BONELL, D.A. GILMOUR & D.S. CASSELLS
A PRELIMINARY SURVEY OF THE HYDRAULIC PROPERTIES OF RAINFOREST SOILS IN TROPICAL NORTH–EAST QUEENSLAND AND THEIR IMPLICATIONS FOR THE RUNOFF PROCESS

A.C. IMESON
STUDIES OF EROSION THRESHOLDS IN SEMI-ARID AREAS. FIELD MEASUREMENTS OF SOIL LOSS AND INFILTRATION IN NORTHERN MOROCCO

J.B. THORNES & A. GILMAN
POTENTIAL AND ACTUAL EROSION AROUND ARCHAEOLOGICAL SITES IN SOUTH EAST SPAIN

D TROPEANO
SOIL EROSION ON VINEYARDS IN THE TERTIARY PIEDMONTESE BASIN (NORTHWESTERN ITALY): STUDIES ON EXPERIMENTAL AREAS

TH.W.J. VAN ASCH
WATER EROSION ON SLOPES IN SOME LAND UNITS IN A MEDITERRANEAN AREA

P VAN DER LINDEN
SOIL EROSION IN CENTRAL–JAVA (INDONESIA). A COMPARATIVE STUDY OF EROSION RATES OBTAINED BY EROSION PLOTS AND CATCHMENT DISCHARGES

W.E.H. CULLING
SLOW PARTICULARATE FLOW IN CONDENSED MEDIA AS AN ESCAPE MECHANISM: I. MEAN TRANSLATION DISTANCE

W.E.H. CULLING
RATE PROCESS THEORY OF GEOMORPHIC SOIL CREEP

CLASSIFICATION OF EUROPEAN DUNE COASTS

E.C.F. Bird, Parkville

Summary

European dune coasts can be classified according to the origin of the sand deposits (e.g. supplied by rivers, eroded from cliffs, or carried in from the sea floor); their composition (e.g. calcareous, quartzose, mixed); their predominant forms (e.g. foredunes, parabolic dunes, transgressive dunes); or their modern dynamics (e.g. drifting dunes, stabilised dunes). Examples are given of each categorie.

1 Introduction

Coastlines can be classified in various ways, but if attention is confined to dune coasts the chief factors to be considered are the source and nature of the sand supply, the degree of exposure to wind and wave action, and the kinds of coastal configuration that permit dune deposition. In this contribution space does not allow presentation of even brief classifications of coastal dunes such as presented by OLSON & VAN DER MEULEN (1989). It rather gives an overview of several distinct coastal features in relation to related dune forms.

The European coastline is intricate and varied in relation to geology (STEERS 1982, KLIJN, this volume), with rock types that range from the older and more resistant formations that outcrop as promontories and headlands along the Atlantic seaboard to the soft and often rapidly retreating coastlines on unconsolidated sedimentary formations, including glacial drift eposits, notably around the North Sea and the southern Baltic. Intermediate are the outcrops of chalk and sandstone which form distinctive cliffs and shore platforms along the shores of the English Channel.

2 Provenance of dune sands

Coastal dunes are derived from sandy beaches by the action of onshore winds as shown by KLIJN (this volume). The beaches are found on sectors where sand has been delivered to the coast by rivers, derived from the erosion of arenaceous cliffs and carried alongshore, or washed in from the sea floor. Sand dunes are formed where there is a sufficient supply of sand, where onshore winds are strong enough to move it, and where there is an area, generally low-lying, where it can accumulate. Dunes are well developed behind wide sandy beaches, especially on coasts where there is a large tide range, and a broad sandy foreshore is exposed at low tide (photo 1).

Photo 1: *Dunes formed where onshore winds carry sand from a wide sandy beach exposed at low tide near Pendine on the South Wales coast. Photo by Eric Bird.*

Few European beaches have been fluvially nourished, because most of the rivers flow into drowned valley mouths (estuaries, rias or fiords) which are still being infilled with river sediment, but sand has been delivered to the coast by such rivers as the Guadalquivir in Spain, the Loire in France, and the Tees in England. In each case the fluvial sand supply has been re-worked by wave action and delivered to nearby beaches, which in turn have supplied the dunes of the Gulf of Cadiz (VANNEY et al. 1979), the Bay of St. Nazaire, and the Cleveland coast respectively. In the Mediterranean, dunes have formed behind beaches supplied with fluvial sand on the shores of the Ebro, the Rhône, and the Po deltas.

Derivation of beach and dune sands from eroding cliffs has also been comparatively rare in Europe. Sectors of cliff exposing soft Tertiary and Quaternary sand deposits occur locally in southeastern England for example, where their erosion has yielded sands to the beaches of Bournemouth Bay and the Holderness coast, but these are typically narrow beaches in front of retreating cliffs, and their contributions to dune development have been minor.

Sand from the sea floor has been a major contribution to beaches and dunes on the European coastline (KLIJN, this volume). During low sea level phases in the Pleistocene (correlated with colder climatic conditions and the growth of glaciers and ice sheets) much of the continental shelf bordering Europe was subaerially exposed. Rock outcrops were weathered, and rivers and glaciers deposited an array of sediments, including sandy material, on the emerged sea floor. Some of this material was winnowed and carried inland by onshore

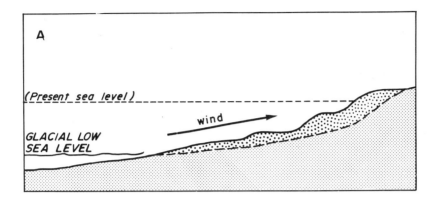

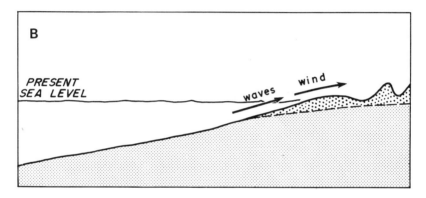

Fig. 1: *Sand movement from the sea floor to beaches and dunes.*
A - during low sea level phases of the Pleistocene, when sand was winnowed from the emerged sea floor;
B - during and since the ensuing marine transgression, when sand was carried shoreward by wave action.

winds, to be deposited as dunes in areas that are now on and behind the coastline (fig. 1,A). The extensive sandy deposits on the southern shores of the Bay of Biscay may have developed in this way, onshore winds generating a supply of aeolian sand that accumulated in Les Landes as massive transgressive dunes, with crests up to 100 metres above sea level, spilling inland at up to 2 metres per year (BRESSOLIER et al., this volume, BUFFAULT 1942).

During intervening milder phases of the Pleistocene, marine transgressions collected sandy sediment and carried it shoreward, and as each transgression slackned this was deposited to form extensive beaches, from which dunes were derived (fig. 1,B). Thus the Late Quaternary marine transgression, which began about 18,000 years ago, and brought the sea to approximately its present level about 6,000 years ago, not only created the essential outline of the European coast, with its drowned valley mouths and submerged lowlands as bays between headlands and promontories of higher ground, but also delivered large quantities of sandy material to build beaches and dunes. Subsequently, with the sea at its present level, the seaward margins of some coastal dune areas have been trimmed back, some of the sand being blown landward, as on the Picardy coast in northern France (ZENKOVICH 1967, 607).

It is noteworthy that the extensive Pleistocene dune formations found on the coasts of the southern Mediterranean, and around the southern continents, especially in Australia, (BIRD 1984, 193) have few counterparts in northern and western Europe. This may be the outcome of dispersal of older dune systems by glacial and periglacial processes during the Last Glacial phase of the Pleistocene. Dissected remnants of Pleistocene dunes persist inland in some areas, such as Picardy, while around the subsiding margins of the North Sea they may be found as buried aeolian formations in the coastal stratigraphy. In southern Europe, Pleistocene dunes occur on the coast near Anzio, in Italy, as a zone of red dunes separated by the Lago di Paola from the Holocene dunes of the coastal fringe (ZEUNER 1959, 233). In general, however, European coastal dunes were largely formed during and after the last (Flandrian) marine transgression, and are therefore of Holocene age.

This was the origin of the sand washed into many bays on the coasts of Brittany to form beaches backed by dunes (HALLEGOUET 1987) and the same process yielded the beaches and dunes of St. Ives Bay, Perran Bay, and Padstow Bay in Cornwall, as well as Braunton Burrows and the Croyde Bay dunes in north-west Devon (WILLIS et al. 1959). In each of these examples, the sands carried in from the sea floor have been augmented by relatively small amounts of sand derived from weathered formations mantling headland cliffs, notably the rubbly material known as Head on the Cornish coast.

In some sectors shoreward drifting continues, as in Carmarthen Bay, South Wales, where sand from nearshore shoals is still prograding the beach, and nourishing the dunes at Pembrey. Shoals in Cardigan Bay have been the source of sand drifting shoreward to the beaches and dunes of Tremadoc Bay, while the dunes of Newborough Warren in south-eastern Anglesey, received sand that moved in from shoals in Carnarvon Bay.

On the Atlantic coast, beach and dune sands supplied from the sea floor generally contain proportions of calcareous sediment, notably comminuted marine shells washed in from the continental shelf. This is the origin of the shelly beaches on the coasts of Scotland, from which calcareous dunes, and the sand sheets known as machair in the Hebrides, have been derived (MATHER 1979). On the other hand, the sand washed in to form the beach at Studland, backed by the dunes of South Haven Peninsula in Dorset, on the south coast of England, is strongly quartzose, having been derived by wave working of weathered Tertiary terrigenous sandstones on the floor of Bournemouth Bay.

On the Mediterranean coasts of Europe, sand washed in from the sea floor has formed the beaches and dunes of Languedoc in south-west France, and the dune-capped barrier islands of the northern Adriatic, the east coast of Corsica, and the shores of Sardinia.

Around the Irish Sea, the North Sea and the Baltic, glacial drift deposits submerged by the sea have been re-worked by wave action, and sand sorted from them has been deposited on the bordering coasts. This is the origin of much of

Photo 2: *Dunes formed on the Baltic coast near Kalajoki, Finland, where wave action has reworked sandy esker deposits. Photo by Eric Bird.*

the sand in the beaches and dunes north of Liverpool (GRESSWELL 1953), and on the northern end of the Isle of Man. Re-worked glacial drift provided the sand for the beaches that border the southern North Sea from Belgium to western Jutland, including the dunes on the barrier islands from Texel in the Netherlands around to Rømø and Fanø in Denmark. The latter are the upper parts of massive sand formations built upward during the Late Quaternary marine transgression on the seaward side of the Wadden Sea, now an area of muddy sedimentation and marshlands (VAN STRAATEN 1961, 1965).

The relationship between beaches and source area of sandy glacial drift is clearly shown in southern Norway, where sandy beaches and dunes occur where morainic drift zones cross the coastline, or where they existed just offshore (KLEMSDAL 1959). On the coasts of the Baltic Sea, many beach and dune areas are found where sandy eskers have been submerged and re-worked by wave action, as in the vicinity of Oulu in Finland (ALESTALO 1979) (photo 2), or where sandy deposits eroded out of cliffs cut in glacial drift have been carried alongshore and deposited, for example on the Łeba spit in Poland (KLIEWE 1973, BORÓWKA 1980). In eastern England, Blakeney Point and Scolt Head Island occupy sectors where Last Glacial morainic deposits intersected the Norfolk coastline, yielding gravels for shingle beach construction and sand which form wide beaches and derived dunes. The dunes of the Lincolnshire coast,

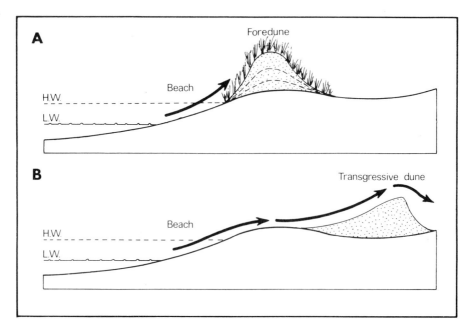

Fig. 2: *A - formation of a foredune where vegetation traps sand blown to the back of a beach; B - formation of transgressive dunes, where sand blown to the back of a beach is not trapped by vegetation.*

near Skegness and Mablethorpe, are also the product of wave re-working of sand from shoals derived from glacial drift, as are the Holy Island dunes on the coast of Northumberland, while in north-east Scotland Culbin Sands and the Rattray Head dunes had a similar origin from sandy morainic zones (STEERS 1937).

Dunes have not developed on steep sectors of the coastline, such as the granitic promontory of Land's End in south-west England, the cliffy headlands of Britanny, or the bold limestone coasts of the Mediterranean. They do not occur where there is active recession of cliffs, as on the chalk coasts bordering the English Channel, or the retreating cliffs of glacial drift in East Anglia and the Danish archipelago (SCHOU 1945). In general, they are poorly developed within deep inlets, such as the rias of Calabria in north-west Spain or the fiords of western Norway. They are also missing from sheltered subsiding areas, such as the Thames estuary and the Essex coast, where mudflats and marshlands predominate. On the coasts of the Baltic, beaches and dunes are not found on emerging rocky promontories, and in the absence of a sand source the shores are marshy.

3 Coastal dune morphology

Dunes built up behind sandy beaches typically form a ridge, known as a foredune (fig. 2,A), colonised and stabilised by grassy and shrubby vegetation. Marram grass (*Ammophila arenaria*), sea wheat grass (*Agropyron junceum*) and sea

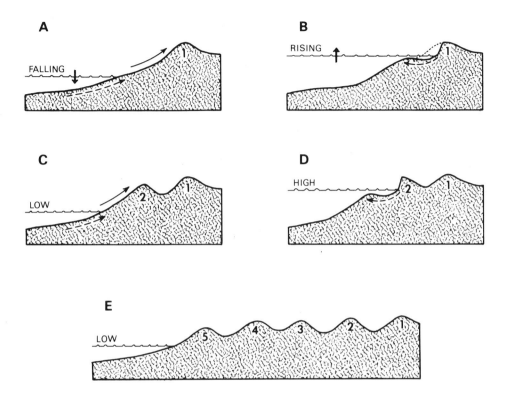

Fig. 3: *Development of successive parallel dune ridges as the result of oscillations of sea level on a coast receiving a sand supply.*
A - during an episode of falling sea level sand is carried on to the beach to build a foredune;
B - when sea level rises, the seward margin of the foredune is cut back;
C - during the next phase of falling sea level sand accretion develops a second foredune seaward of the first;
D - this in turn is cliffed as a sea level rise;
E - continued alternations produce a succession of parallel ridges.

rocket (*Cakile maritima*) are common pioneer species, along with lyme grass (*Elymus arenarius*) in northern Europe, and there is succession to scrub dominated by such shrubs as sea buckthorn (*Hippophäe rhamnoides*) (RANWELL 1972, BRESSOLIER & THOMAS 1977). On some sectors, as at Sea Palling in northeast Norfolk, there is only one foredune, but where the coastline has prograded as the result of sand accumulation, there is often a series of successively formed parallel foredunes (primary dune ridges) and sometimes cut-off back plains (primary dune slacks), as at Tentsmuir on the Scottish coast, the Magilligan Foreland in Northern Ireland (CARTER & WILSON 1988, CARTER, this volume, WILSON, this volume), and the Darss Foreland in the German Democratic Republic (GUILCHER 1978). The formation of parallel dune ridges and related dune slacks is the outcome of cut-and-fill, alternations of coastal erosion and accre-

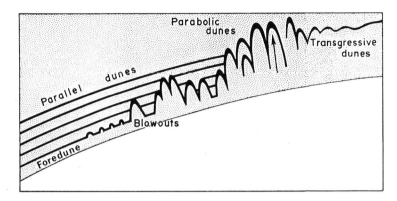

Fig. 4: *Parallel dunes, developed as successive foredunes on a prograding coast, are held in place by a vegetation cover. Where this is interrupted, blowouts develop, and these can grow into parabolic dunes. With further depletion of vegetation, transgressive dune ridges may be formed, drifting inland.*

tion produced either by phases of storm wave erosion interrupting calm-weather accretion (BIRD 1984, 180), minor oscillations of sea level (fig. 3), or intermittent growth of spits or sandy barrier islands, such as "wadden islands". The height and spacing of the parallel dunes is a function of the rate of sand supply, the frequency of 'cut-and-fill' alternations, and the effectiveness of vegetation in trapping blown sand.

The next categories of coastal dunes are in northwestern European regions often called secondary dunes (VAN DIEREN 1934).

Where the vegetation cover has been locally breached as the result of damage to vegetation by wave erosion or by clearing, grazing, burning or trampling, onshore winds have developed blowouts, hollows cut through the foredune, with a nose of sand spilling inland (fig. 4). The growth of such blowouts may eventually produce elongated parabolic dunes, with trailing ridges of sand held in place by vegetation as the nose advances, along axes related to the onshore wind resultant (LANDSBERG 1956, JENNINGS 1957). Examples of these are found in the Sands of Forvie on the east coast of Scotland (ROBERTSON-RINTOUIL & RITCHIE, this volume). Repeated disruption, alternating with phases of natural (or artificial) revegetation produces an irregular hummocky dune topography, as in the Perran Sands in Cornwall.

With further depletion of vegetation, coastal dunes may become generally mobile and transgressive (fig. 4), advancing inland as one or more ridges of drifting sand. At Newborough Warren in Anglesey three such transgressive ridges moved in from the coastline, separated by low corridors which show stages in vegetation succession related to the dune advance (RANWELL 1958). Several massive sand waves were migrating inland in the Landes of Gascony before they were stabilised by the planting of vegetation they now carry pine forests (BUFFAULT 1942, BRESSOLIER et al., this volume). Where unvegetated sand dunes persist, they may include barchans similar to those found in deserts (GRIPP 1961, TSOAR, this volume).

Parabolic dunes and transgressive dunes are not necessarily derived from foredunes and parallel dune systems. Where the backshore vegetation is weak, the rate of sand supply very rapid, or onshore winds very strong, the initial backshore dunes may develop either as parabolic dunes with patches of vegetation holding the trailing ridges, or as transgressive sand dunes (fig. 2,B). This is commoner on semi-arid and arid coasts, but there are European examples, notably on the coast of the Desert des Agriates in Corsica, where sand washed up on the beach has blown directly inland as transgressive dunes.

4 Conclusion

European coastal dunes are thus varied in form and origin. Most have been influenced to some extent by human activities, either by depleting the natural vegetation cover and mobilising dunes that were previously stable, or by deliberately planting vegetation to arrest drifting sand. Nevertheless, there are examples of each of the categories of coastal dune landforms that have been identified on the global scale (ZENKOVICH 1967, NORRMAN 1981).

References

ALESTALO, J. (1979): Land uplift and development of the littoral and eolian morphology on Hailuoto, Finland. Acta University of Oulu Geology **3**, 109–120.

BIRD, E.C.F. (1984): Coasts. Blackwell, Oxford.

BRESSOLIER, C. & THOMAS, Y.F. (1977): Studies on wind and plant interactions on French Atlantic coastal dunes. Journal of Sedimentary Petrology **47**, 331–338.

BRESSOLIER, C., FROIDEFOND, J.M. & THOMAS, Y.F. (this volume): Dune chronology along the southwestern coast of France.

BORÓWKA, R.K. (1980): Present day dune processes and dune morphology on the Łeba barrier, Polish Baltic coast. Geografiska Annaler, Series **A**, **62**, 75–82.

BUFFAULT, P. (1942): Histoire des dunes maritimes de la Gascogne. Bordeaux, 446 p.

CARTER, R.W.G. (this volume): The geomorphology of coastal dunes in Ireland.

CARTER, R.W.G. & WILSON, P. (1988): Geomorphological, sedimentological and pedological influences on coastal dune development in Ireland. Journal of Coastal Research, Speciel Issue **3**, 27–31.

DIEREN, J.W. VAN (1934): Organogene Dünenbildung. Thesis, Den Haag, Nijhoff.

GRESSWELL, R.K. (1953): Sandy shores in South Lancashire. University of Liverpool.

GRIPP, K. (1961): Über Werden und Vergehen von Barchanen an der Nordsee-Küste Schleswig-Holsteins. Zeitschrift für Geomorphologie **5**, 24–36.

GUILCHER, A. (1978): Les crêtes littorales dunifiés de type Darss de la baie de Goulven, Bretagne comparées à celles de l'île de Wolin, Pologne. Wissenschaft Zeitschrift der Ernst-Moritz-Arndt-Universität Greifswald **27**, 26–33.

HALLEGOUET, B. (1978): L'évolution des massifs dunaires du Pays de Léon, Finistère. Penn ar Bed **95**, 417–430.

JENNINGS, J.N. (1957): On the orientation of parabolic or U-dunes. Geographical Journal **123**, 474–480.

KLEMSDAL, T. (1969): Eolian forms in parts of Norway. Norsk Geografisk Tidsskrift **23**, 49–66.

KLIEWE, H. (1973): Zur Genese der Dünen im Küstenraum der DDR. Petermanns Geographische Mitteilungen **117**, 161–168.

KLIJN, J.A. (this volume): Dune forming factors in a geographical context.

LANDSBERG, S.Y. (1956): The orientation of dunes in Britain and Denmark in relation to the wind. Geographical Journal **122**, 176–189.

MATHER, A.S. (1979): Physiography and management of coastal dune systems in the Scottish highlands and islands. In: A. Guilcher (ed.), Les Côtes Atlantiques de l'Europe. Publications du Centre National pour l'exploitation des Océans. Actes de Colloques **9**, 251–260.

NORRMAN, J.O. (1981): Coastal dune systems. In: E.C.F. Bird & K. Koike (eds.), Coastal Dynamics and Scientific Sites. Komazawa University, Tokyo, 119–157.

OLSON, Y.S. & VAN DER MAAREL, E. (1989): Coastal dunes in Europe; a global view. In: F. van der Meulen, P.D. Jungerius & J-Visser (eds.), Prspectives in coastal dune management. 3–32. SPB Acad. Publ., The Hague.

RANWELL, D.S. (1958): Movement of vegetated sand dunes at Newborough Warren, Anglesey. Journal of Ecology **46**, 83–100.

RANWELL, D.S. (1972): The ecology of salt marshes and sand dunes. Chapman Hall, London.

SCHOU, A. (1945): Det marine foreland. Folia Geographica Danica **4**, 236 p.

ROBERTSON-RINTOUIL, M. & RITCHIE, W. (this volume): The geomorphology of coastal sand dunes in Scotland. A review.

STEERS, J.A. (1937): The Culbin Sands on Burghead Bay. Geographical Journal **90**, 498–528.

STEERS, J.A. (1982): Europe, coastal morphology. In: M.L. Schwartz (ed.), The Encyclopaedia of Beaches and Coastal Environments. Hutchinson & Ross, New York. 409–420.

STRAATEN, L.M.J.U. VAN (1961): Directional effects of winds, waves and currents along the Dutch North Sea coast. Geologie en Mijnbouw **40**, 333–346.

STRAATEN, L.M.J.U. VAN (1965): Coastal barrier deposits in south and north Holland, in particular the area around Scheveningen and Ijmuiden. Mededelingen van de Geologische Stichting, N.S. **17**, 41–75.

TSOAR, H. (this volume): Trends in the development of sand dunes along the southeastern Mediterranean coast.

VANNEY, J.R., MENANTEAU, L. & ZAZO, c. (1979): Physiographie et évolution des dunes de Basse Andalousie. In: A. Guilcher (ed.), Les côtes Atlantiques de l'europe. Publications du Centre National pour l'exploitation des Océans. Actges de Colloques **9**, 277–286.

WILLIS, A.J., FOLKES, B.F. HOPE-SIMPSON, J.F. & YEMM, E.W. (1959): Braunton Burrows: the dune system and its vegetation. Journal of Ecology **47**, 1–24 and 249–288.

WILSON, P. (this volume): Coastal dune chronology in the north of Ireland.

ZENKOVICH, V.P. (1967): Processes of Coastal Development (ed. J.A. Steers). Oliver and Boyd, New York.

ZEUNER, F.E. (1959): The Pleistocene period. Hutchinson, London.

Address of author:
Eric C.F. Bird
Department of Geography
University of Melbourne
Parkville, Australia 3052

COASTAL DUNES IN POLAND

R.K. **Borówka**, Poznan

Summary

On the Polish Baltic coast are very well developed complexes of foredunes and also fields of present-day migrating parabolic and barchan dunes. The complexes of foredunes which appear especially on the Świna Barrier and Hel Spit were formed during the last 5000 years. The development of the migrating dune complexes was multiphase. For the last 4000 years four stages of eolian activity are recognized. The periodically repeated stages of an intense revival of eolian processes are closely related to man's economic activity.

1 Introduction

The Polish Baltic coast consists of sandy parts and cliff sections forming a smooth abrasive-accumulative coastline (fig. 1). As a rule the cliff parts of the coast cut into ground moraine plateaux and end moraine hills consisting primarily of glacial tills and fluvioglacial sands and gravel. The sandy parts of the coast, usually in the form of barriers with coastal dunes are found close to the Late Glacial depressions and valleys separating coastal lakes, bays and swamp plains from the sea.

ISSN 0722-0723
ISBN 3-923381-23-9
©1990 by CATENA VERLAG,
D-3302 Cremlingen-Destedt, W. Germany
3-923381-23-9/90/5011851/US$ 2.00 + 0.25

The most decisive influence on the development of the southern Baltic coastline was exerted by the Littorina transgression during the Atlantic period. At the initial phase of the transgression (8000–7500 years BP) the existing coastline was situated about 31 m below the present sea level (ROSA 1987, 151) and usually extended from several to a few dozen kilometres northwards from the present coastline (fig. 1). At the maximum rate of the Littorina transgression (7500–5500 years BP) the barriers shifted southwards following the relative rise of the sea level. This is confirmed by the fact that the sandy barriers with dunes are usually underlain by peats and lagoon or lacustrine deposits (BRODNIEWICZ & ROSA 1967, 64, ROSA & WYPYCH 1980, 39, BORÓWKA & ROTNICKI 1988, 27). In some places (east of Ustka) coastal dunes connected with the barriers entered the ground moraine plateaux, forming at present cliff-top dunes.

On the sandy barriers are well developed complexes of coastal dunes that display distinctive morphologic characteristics. Among these complexes, there are:

- complexes of foredunes (the Świna Barrier and the eastern part of the Hel Spit),

- single ridges of foredunes, which appear in a very narrow part of sandy barriers (the middle part of the Polish coast between Mielno and

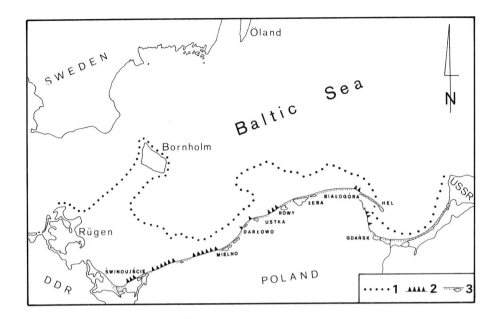

Fig. 1: *Polish Baltic coast.*
1 - extent of the coastline at the initial phase of the Littorina transgression; 2 - cliff coast; 3 - sandy coast with coastal dunes.

Darłowo, the western part of the Hel Spit, the Vistula Barrier — east of Gdansk),

- complexes of parabolic and barchan dunes (west of Ustka, the Łeba Barrier between Rowy and Białogóra).

The most interesting coastal dune fields occur on the Świna Barrier (fig. 2) and Łeba Barrier (fig. 3).

2 The Świna Barrier dunes

The Świna Barrier dunes is an area where numerous dune ridges to be interpreted as a sequence of foredunes have been formed. Their origin is closely related to rapid coastal accretion and the development of spits in that area. KEILHACK (1912, 215) distinguishes between three generations of ridges that were formed as foredunes, the so-called brown, yellow and white dunes, which differ in size and degree of soil cover development.

The meridionally oriented brown dunes from 2 to 8 m in relative height (fig. 2) are covered by very well developed podzols with a brown illuvial horizon, partially cemented by iron oxides. Particular ridges of the brown dunes are separated by peat-filled troughs. The radiocarbon dates on bottom peat deposits, as well as palynological data indicate that the brown dunes were formed during the Subboreal and the earlier part of the Subatlantic Period, at about 5000 to 1800 years BP (PRUSINKIEWICZ & NORYŚKIEWICZ 1966, 85).

The parallel-oriented yellow dunes from 6 to 10 m in relative height (fig. 2) are covered by weakly developed podzols with a yellow illuvial horizon. These dunes were formed between the fifth century A.D. and the midsev-

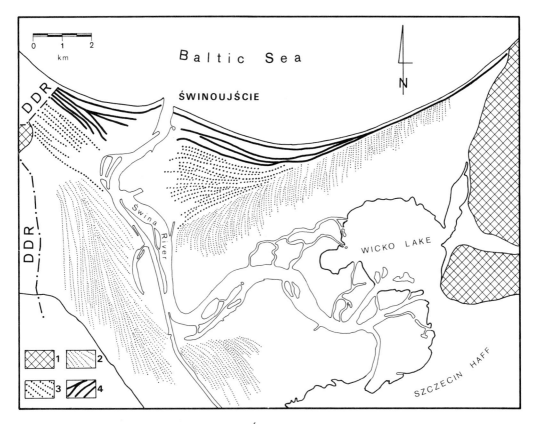

Fig. 2: *Geomorphology of the Świna Barrier (after Keilhack 1912).*
1 - end moraines; 2 - brown dunes; 3 - yellow dunes ; 4 - white dunes.

enteenth century (PRUSINKIEWICZ & NORYŚKIEWICZ 1966, 85).

The parallel-oriented white dunes attain a maximum relative height of 10 to 20 m. KEILHACK (1912, 218) estimated their age on the study of Swedish maps of 1694. In his opinion, their deposition began in the A.D. mid-seventeenth century.

3 The Łeba Barrier dunes

The Łeba Barrier is presently developing under the influence of intense eolian processes. It contains migrating dunes shifting under the influence of the prevailing westerly winds, as well as fixed dunes covered by vegetation, sometimes artificially introduced (fig. 3).

The Łeba Barrier includes the following types of dune forms (MISZALSKI 1973, 139): barchans (17%), barchan-arc dunes (15%) and arc or parabolic dunes (68%). The relative height of these forms is from several metres to the maximum of about 56 m. Such dunes, partially or completely lacking vegetation, migrate eastwards at an average rate of 1 to 10 meter per year (MISZALSKI 1973, 112). In a narrow zone behind the beach complexes of former foredune ridges occur

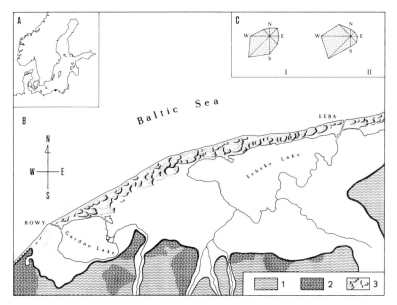

Fig. 3: *Location (A), geomorphic setting of the Łeba Barrier (B) and distribution of frequency (C-I) and effectiveness (C-II) of wind blowing from particular directions at Łeba.*

1 - moraine plateau; 2 - end moraines; 3 - coastal dunes.

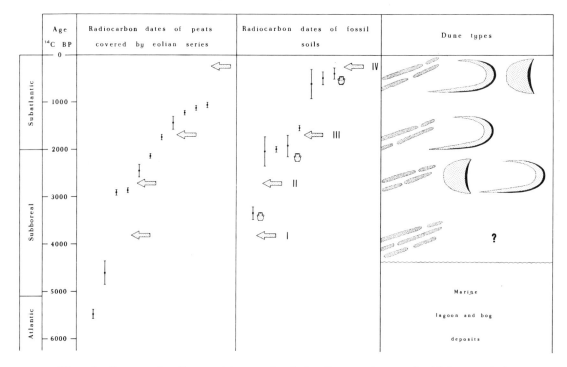

Fig. 4: *Stages of eolian activity on the Łeba Barrier during the Holocene (after BORÓWKA, in press).*

at heights of 10–12 m above sea level. In some parts of the barrier the foredunes have been eroded by the sea. Farther inland "fossil" dune cliffs are often separated from the shoreline by younger dune ridges. Between specific dunes, it is also possible to find deflation forms indlucing not only large inter-dune troughs but also smaller basins or deflation hollows, as well as deflation remnants usually in the shape of elongated WNW-ESE oriented ridges. The deflation aras contain outcrops of fossil soil horizons which cover the old fossil dune forms.

Analysis of the lithostratigraphic, chronostratigraphic (radiocarbon dating) and geomorphological positions of fossil soils and eolian deposits serves as the basis of reconstruction of dune processes history on the Łeba Barrier (BORÓWKA, in press). Four stages of eolian acitvity can be recognized for the last 4000 years (fig. 4).

At the earliest stage of dune formation (about 4000–3500 years BP) low forms of foredune type developed. They are covered by fossil soils containing numerous pieces of charcoal and also fragments of ceramics and well preserved vessels dating from the late Bronze Age (BORÓWKA & TOBOLSKI 1979, 29, BORÓWKA, in press).

The second stage of eolian activity (about 3300–2500 years BP) on the Łeba Barrier brought about formation of not only foredunes but also barchans and parabolic dunes which migrated under the influence of westerly winds. These old migrating dunes are covered by fossil podzols with a very well developed brown illuvial horizon. These podzols are radiocarbon dated to about 2000 years BP (KOBENDZINA 1968, 75, TOBOLSKI 1975, 60, BORÓWKA & TOBOLSKI 1979, 25). The fossil soils were formed when oak forest grew (TOBOLSKI 1975, 46).

The most marked imprints of the third stage of eolian processes occur in the vicinity of the north-western margin of Łebsko Lake although their traces are detectable in the middle part of the Łeba Barrier. Crescent and parabolic dunes reaching about 20–25 m above sea level in height are associated with this stage. Soils developed on these dunes, on which the radiocarbon date of 1540 ± 50 years BP was recorded, display slightly different morphologic characteristics, compared with the podsols dated to about 2000 years BP (TOBOLSKI 1975, 38, BORÓWKA 1975, 41). They contain a poorly developed illuvial horizon. The pollen analysis shows that they were largely formed under the canopy of a pine forest (TOBOLSKI 1975, 50).

Between 1500 and 500 years BP the eolian activity came to a halt. Its revival is dated archaeologically and by the radiocarbon datings method to more or less the XVth Century. Since then permanent and intense development of dune processes has been observed.

It is characteristic that almost all horizons of fossil soils found in the Łeba Barrier area are accompanied by traces of man's impact. The fact that fossil soils and peat deposits fairly frequently contain pieces of charcoal, as well as archaeological finds and documentary evidence suggest that periodically repeated stages of an intense revival of eolian processes on the Łeba dunes are closely related to man's economic activity which has lasted there for the last three milennia. The environment of coastal dunes is one of the least resistant and thus each kind of man's activity may produce considerable changes in veetation and dune morphology.

References

BORÓWKA, R.K. (1975): Problem of the morphology of fossil dune forms on the Łeba Bar. Quaestiones Geographicae **2**, 39–51.

BORÓWKA, R.K. & TOBOLSKI, K. (1979): New archaeological sites on the Łeba Bar and their significance for paleogeography of this area (in Polish). Badania Fizjograficzne nad Polska Zachodnia **32A**, 21–29.

BORÓWKA, R.K. & ROTNICKI, K. (1988): New data on the geologic structure of the Łeba Barrier in Polish. Poznańskie Towarzystwo Przyjaciół Nauk, Sprawozdania Wydziału Matematyczno-Przyrodniczego **105**, 26–29.

BORÓWKA, R.K. (in press): The Holocene development and present morphology of the Łeba Dunes, Baltic Coast of Poland. In: K.F. Nordstrom, N.P. Psuty & R.W.G. Carter (eds.), Coastal Dunes: Processes and Morphology. John Wiley and Sons Ltd., Chichester.

BRODNIEWICZ, I. & ROSA, B. (1967): The boring hole and the fauna at Czołpino, Poland. Baltica **3**, 61–86.

KEILHACK, K. (1912): Die Verlandung der Swinepforte. Jahrbuch der Königlichen Preussischen Geologischen Landesanstalt **32**, 2, 209–244.

KOBENDZINA, J. (1968): Wydmy Słowińskiego Parku Narodowego. Ziemia **1967**, 70–80.

MISZALSKI, J. (1973): Present-day eolian processes on the Slovinian Coastline; a study of photointerpretation (in Polish). Dokumentacja Geograficzna **3**, 1, 1–150.

PRUSINKIEWICZ, Z. & NORYŚKIEWICZ, B. (1966): Problem of the age of podzols on brown dunes of bay bars of river Świna in the light of a palynological analysis and dating by radiocarbon C-14 (in Polish). Zeszyty Naukowe Uniwersytetu Mikołaja Kopernika w Toruniu, Nauki Matematyczno-Przyrodnidcze, Geografia **5**, 75–88.

ROSA, B. & WYPYCH, K. (1980): O mierzejach wybrzeża południo wobałtyckiego. Peribalticum **1**, 31–44.

ROSA, B. (1987): Pokrywa osadowa i rzeżba dna. Bałtyk Południowy. (ed. B. Augustowski). Zakład Narodowy Imienia Ossolińskich, Gdańsk, 75–172.

TOBOLSKI, K. (1975): Palinological study of fossil soils on the Łeba Bar in the Słowiński National Park (in Polish). Poznańskie Towarzystwo Przyjaciół Nauk, Prace Komisji Biologicznej **41**, 1–76.

Address of author:
Ryszard K. Borówka
Quaternary Research Institute
Adam Mickiewicz University
61-701 Poznań
Fredry 10
Poland

THE GEOMORPHOLOGY OF COASTAL DUNES IN IRELAND

R.W.G. **Carter**, Coleraine

Summary

The coastal dunes of Ireland exhibit wide regional contrasts related to a variety of environmental gradients, embracing both processes and sediments: geomorphology varies according to not only location but the rate of sediment input. Dunes with relatively high inputs but no outputs prograde; those with high inputs and high outputs are unstable, often unvegetated systems. Where sand supply is low, dunes suffer marine erosion, especially where sediment is dispersed inland, often via blowouts. Transitional dunes appear to remain stable, with balanced exchanges between land and sea. In these circumstances foredunes are rare.

1 Introduction

The coast of Ireland includes about 1000 km (20%) of dunes covering an area of over 160 km². Almost all these dunes have formed within the last 6000 years as a response to fluctuations in sea-level and sediment supply (CARTER et al. 1989a). The Irish coastal dunes embrace a number of distinct types (fig. 1), largely associated with local and regional environmental differences — sea level, sediment supply, waves, tides, vegetation and man's activities. These factors have varied through time leading to phases of stability and instability within many dunes systems (see WILSON, this volume). Most of the Irish foredunes are fixed by sand binding and sand stabilising vegetation, with the most important plants being *Ammophila arenaria, Elymus farctus* and in places *E. arenaria*. Inland the dune successions are often truncated and indistinct. RILEY (1976) suggested that the two dominant environmental gradients are associated with

i) salinity (i.e. exposure and distance from the sea) and

ii) grazing by animals (rabbits, cattle and sheep).

Many dunes systems are a mosaic of stable and unstable surfaces. In southeast and northeast Ireland there is a number of sites, where despite the presence of dune vegetation, eolian sediments are largely absent. These 'dunes' are primarily marine or alluvial sediments which have been colonised by arenaceous or rudaceous coastal halophytes.

There have been very few studies of dune geomorphology **per se**, although physiographic aspects have been discussed by WILCOCK (1976), JEFFREY (1977), QUINN (1977) among others. The aim of this short paper is to synthesise some of the existing informa-

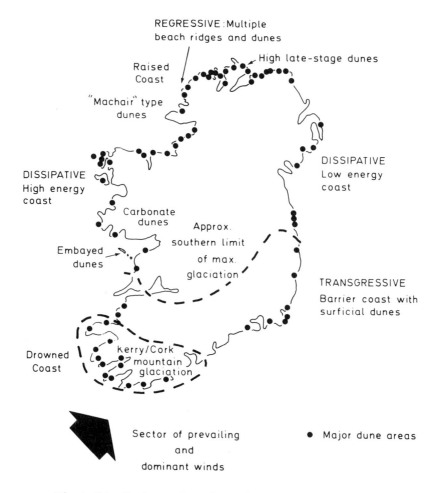

Fig. 1: *Distribution and typology of coastal dunes in Ireland.*

tion from a geomorphological perspective, and highlight some of the distinctivenesses encountered in the Irish dune systems.

2 Background

For a relatively small island, Ireland harbours considerable contrasts in its coastline. The west coast is dominated by high energy incident waves and onshore winds, with most beach and dune sediments contained between headlands and within rock-bounded estuaries. In comparison, the east coast bordering the shallow epi-continental Irish Sea, comprises long stretches of eroding clifflines interspersed with beach ridge plains and dunes. Although ocean swell waves retain an influence, the east coast is dominated by frequent storm waves (ORFORD 1989). The north and south coasts represent transitions between the west and east extremes, although they have experienced a marked difference in

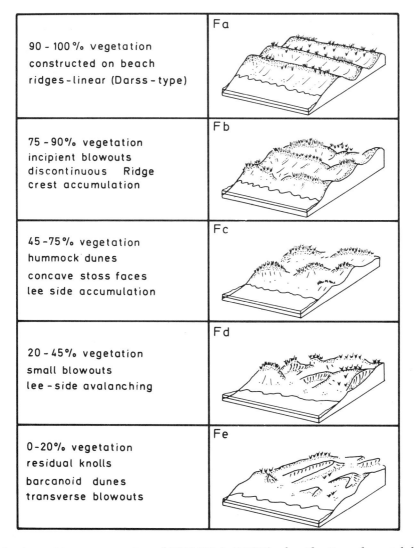

Fig. 2: *A graphic representation of SHORT & HESP's classification of coastal dunes, based on shoreline morphodynamics.*

sea level history. In the south, sea level has risen inexorably, although at a decellerating rate throughout the Holocene, while in the north sea level peaked around 4500 to 6500 years BP (CARTER et al. 1989a), followed by a fall and a subsequent rise after. These variations in sea level have been very important in controlling the evolution of coastal dunes.

Most Irish dunes fall into categories Fb and Fc of SHORT & HESP's (1982) biophysical classification (fig. 2), inasmuch as they retain the basic shore-normal ridge structure under a relatively

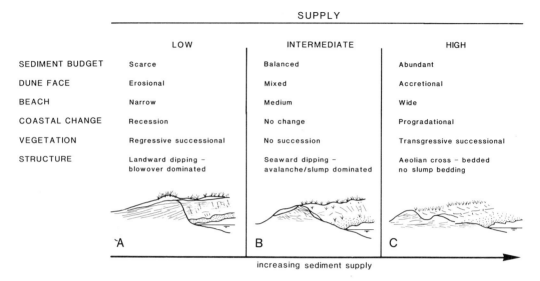

Fig. 3: *Current Irish coastal dune topography is closely associated to the local sediment supply. This figure shows the three basic types associated with low, balanced and high sediment influxes. Another high flux situation is shown in fig. 4.*

effective vegetation cover. However, it is possible to find examples of Fa (Darss type dunes — CARTER 1988) and Fd and Fe (largely unvegetated dunes). At present, almost all Irish dune systems are not receiving large volumes of fresh sand, so that genuine progradational forms are rare. However there is a lot of sediment recycling, on scales ranging from short (weeks, months, years) to long (decades, centuries, millenia), so that geomorphological changes to the shorelines are common. Any variation in landform expression appears to be related to the local recycling rate and its mode of operation. One main threshold associated with recycling is related to the ability of sand-binding vegetation to survive engulfment. Among the common dune species there is a wide range of sand burial tolerances (see CARTER 1988, 324) from a few centimeters for ground cover plants to 0.3 to 0.6 m for sand-binding grasses like *Ammophila*, particularly if deposition coincides with the growing season. However, it should be noted that sand deposition has a significant impact on all manner of ecological and physiological factors associated with dune vegetation. Accordingly, the Irish dunes may be divided into those where sand supply rates are high and those where supply is low (fig. 3), separated by a category where sand is neither abundant or scarce.

3 Dune Geomorphology

3.1 High sediment supply systems

Dune systems with relative high sediment inputs occur where there is a sufficiently dynamic conjunction of waves, tidal and eolian processes to maintain a steady supply of sand sized material. Two basic forms emerge, one an accumu-

Photo 1: *Rapidly prograding dunes at Clonmass in Co. Donegal. These dunes have accumulated in the last two decades as part of readjustment within the Clonmass estuary (see SHAW 1984).*

lative form where sediment is normally available, comprising rapidly extending foredunes, and two, a throughput form where recycling prevails and there is no marked trend in the shoreline.

Accumulative forms are typified by an excess of eolian sand input, leading to the development of beach ridges and ridge-top dunes. Such systems have been described by CARTER (1975) and CARTER & WILSON (1990) from Magilligan in Co. Londonderry, by SHAW (1984) from Clonmass, Co. Donegal and by HARRIS (1974) from North Bull Island in Co. Dublin. In all cases the dune system has prograded rapidly over the last few decades (photo 1), interspersed with occasional phases of retreat. At Magilligan Point, erosion of adjacent shorelines (CARTER 1986, CARTER & STONE 1989) — possibly resulting in part from sea-level rise — has led to the lateral development of beach ridges and thence foredunes. The average growth rate of the foredunes varies between 6 and 9 m^3 m^{-1} a^{-1}, which translates into a vertical accretion of between 0.3 and 0.7 m a^{-1} (CARTER & WILSON 1990). Each foredune ridge develops until such time as it is cut-off from the beach supply — usually through the formation of a new foredune. Geomorphological stabilisation is accompanied by a rapidly increasing floral diversity, decalcification and accumulation of organic soils (WILSON

Photo 2: *Sand corridor within the high flux system at Inch, Co. Kerry. Sediment flux volume precludes vegetation stabilisation, although there are areas of deposition and vegetation growth within the system.*

1987). It is noteworthy that the Magilligan foredunes include gaps, which never seal (CARTER & WILSON 1990), and act as routes for the dispersal of beach sediment inland.

Sequences of rapid foredune development in Ireland are often associated with the closure of estuary mouths brought about by wetland reclamation (CARTER 1988, 523–525). As the tidal prism volume falls so the estuary mouth becomes hydraulically inefficient and closes. This process is accompanied by a re-allocation of estuary mouth and adjacent shoreline deposits, which favours the creation of new foredunes. In these cases small sand-rich 'islands' exist within more general sand-poor "oceans".

The second type of sand-rich dune systems is associated with tidal 'pumps' (CARTER et al. 1989b) and high sediment fluxes. In such examples sediment is moved repeatedly from beach to dune to tidal pass (fig. 4). Much of this activity takes place at the distal extremity of sand 'spits'. The dunes are mobile (photo 2) with major sand corridors, thick flank accumulations, large 'blowover' slip faces and a general paucity of stabilising vegetation (accretion rates ar too high to allow survival). There are many examples of tidal pump dunes in Ireland, including Inch and Rossbehy in Co. Kerry, Tramore in Co. Waterford and Ballyness in Co. Donegal.

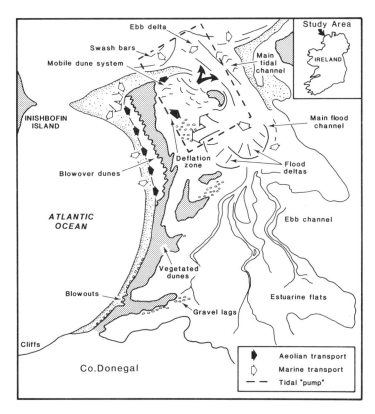

Fig. 4: *The high flux tidal "pump" at Ballyness in Co. Donegal leads to the development of mobile dunes via a coupling of tidal and eolian processes. Gradual dispersal of sediment into estuary sinks is leading to a slow closure of the tidal channel, and a reduction in sediment transport. In time this may allow stabilisation of the dunes.*

3.2 Low sediment supply systems

In Ireland, the lack of an effective sediment supply to dunes is a common occurrence. Most systems are characterised by a slow release of sediment either from marine erosion of following vegetation die-back. At the seward margins of the dunes, marine erosion leads to gradual crestal accumulation (fig. 3A) tapering inland. The net result is the development of a shore-parallel ridge rising to a maximum along the erosional margin. Under strong onshore winds, the interaction of a sharp erosional facet and low sediment supply, may lead to an extensive lateral sand dispersal, creating sand sheets of sub-parallel laminae only one or two grains thick. These sheets often dip landward (at 1 or 2°), forming a machair (BASSETT & CURTIS 1985) surface. A slight variation is often introduced when blowouts are present. These may develop from natural wind gaps, or from marine erosion or human disturbance. Regardless of origin, blowouts serve to funnel material inland often in wide fan-shaped plumes (CARTER et al. 1990).

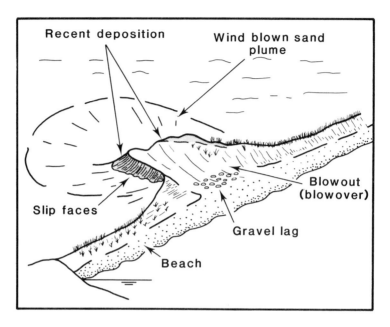

Fig. 5: *As coastal recession proceeds so blowouts serve to disperse sediment landward. Blowouts often accumulate vertically (thus accentuating relief) as sediment passes through. The inner "blowover" deposits comprise both a slipface and a broader plume, the latter often several hundered square metres in extent.*

Maximum deposition in these blowouts occurs on the landward margins, accentuating their overall relief (fig. 5).

Sand-poor systems also display a capacity for reorganisation, usually in relation to dominant winds. In the most prominent case, at Portstewart, Co. Derry, the system has become a series of near-parabolic dunes migrating downwind (WILCOCK 1976). Over time the creation and healing of deflation landforms leads to the development of largely chaotic topography, which assumes neither the shore-parallel expression of coastal dunes, nor the wind-orientated structures of a fully mobile, aeolian-driven system.

3.3 Transitional forms

Perhaps the least spectacular dunescapes are those where sand input is moderate. Under these conditions, dune vegetation is capable of absorbing all additions of wind blown sand. These dunes, exemplified by the systems at Portrush, Co. Antrim (CARTER 1980), operate a balanced seasonal exchange with the beach and nearshore. Shoreline topography is usually dominated by a large single dune ridge (fig. 3B), comprising seaward-dipping sedimentation units and extensive evidence of mass movements ranging from avalanches to small rotational slips (CARTER 1980, CARTER & STONE 1989). Foredunes are usually absent, although small shadow dunes may form in late summer, before material is moved onto the seaward dune ridges. Slight shifts in the long-term sediment economy may result in either erosion (sand-poor tendency) or progradation (sand-rich tendency).

4 Conclusions

The variety of coastal dunes in Ireland may be explained by several environmental factors working in concert. These include the style and nature of Holocene sea-level change, sediment availability, wave and tide processes, wind regime and human activities. Modern geomorphological features can be related to the volume and rate of local sediment transport, on a scale ranging from high to low material fluxes. The dunes must be viewed as an integral part of the overall coastal system, receiving, storing and releasing material in response to environmental change.

References

BASSETT, J.A. & CURTIS, T.G. (1985): The nature and occurrence of sand-dune machair in Ireland. Proceedings of the Royal Irish Academy **85B**, 1–20.

CARTER, R.W.G. (1975): Recent changes in the coastal geomorphology of the Magilligan foreland. Proceedings of the Royal Irish Academy **75B**, 469–497.

CARTER, R.W.G. (1980): Vegetation stabilisation and slope failure on eroding sand dunes. Biological Conservation **18**, 117–122.

CARTER, R.W.G. (1986): The morphodynamics of beach ridge formation, Magilligan, Northern Ireland. Marine Geology **73**, 191–214.

CARTER, R.W.G. (1988): Coastal Environments. Academic Press, London, 617 pp.

CARTER, R.W.G. & STONE, G.W. (1989): Mechanisms associated with the erosion of sand dune cliffs, Magilligan, Northern Ireland. Earth Surface Processes and Landforms **14**, 1–10.

CARTER, R.W.G. & WILSON, P.W. (1990): The geomorphological, ecological and pedological development of coastal foredunes at Magilligan Point, Northern Ireland. In: Nordstrom, K.F., Psuty, N. & Carter, R.W.G (eds.), Coastal Dunes. Wiley, Chichester.

CARTER, R.W.G., DEVOY, R.J.N. & SHAW, J. (1989a): Late-Holocene sea levels in Ireland. Journal of Quaternary Science **4**, 7–24.

CARTER, R.W.G., HESP, P.A. & NORDSTROM, K.F. (1990): Erosional landforms in coastal dunes. In: Nordstrom, K.F., Psuty, N. & Carter, R.W.G. (eds.), Coastal Dunes. Wiley, Chichester.

CARTER, R.W.G., FORBES, D.L., JENNINGS, S., ORFORD, J.D., SHAW, J. & TAYLOR, R.B. (1989b): Barrier and lagoon coast evolution under differing relative sea-level regimes: examples from Ireland and Nova Scotia. Marine Geology **88**, 221–242.

HARRIS, C.R. (1974): The evolution of North Bull Island, Dublin Bay. Scientific Proceedings of the Royal Dublin Society **A5**, 237–252.

JEFFREY, D.W. (Editor) (1977): North Bull Island, Dublin Bay — a modern coastal natural history. Mount Salus Press, Dublin, 158 pp.

ORFORD, J.D. (1989): A review of tides, currents and waves in the Irish Sea. In: J. Sweeney (Ed.), The Irish Sea — A Resource at Risk. Geographical Society of Ireland, Special Publication no. **3**, Dublin, 18–46.

QUINN, A.C.M. (1977): Sand Dunes — Formation, Erosion and Maintenance. An Foras Forbartha, Dublin, 92 pp.

RILEY, D.H. (1976): Magilligan Dunes, County Londonderry. Classification of the vegetation. Unpublished D.Phil thesis, The New University of Ulster, Coleraine.

SHAW, J. (1984): Clonmass Estuary. In: P. Wilson & R.W.G. Carter (eds.), Northeast Co. Donegal and northwest Co. Londonderry. Irish Association for Quaternary Studies, Field Guide, Number 7, 10–21.

SHORT, A.D. & HESP, P.A. (1982): Wave, beach and dune interactions in southeastern Australia. Marine Geology **48**, 259–284.

WILCOCK, F.A. (1976): Dune physiography and the impact of recreation on the north coast of Ireland. Unpublished D.Phil. thesis, The New University of Ulster, Coleraine.

WILSON, P. (1987): Soil formation on coastal beach and dune sands at Magilligan Point Nature Reserve, Co. Londonderry. Irish Geography **20**, 43–49.

Address of author:
R.W.G. Carter
University of Ulster
Coleraine
Co. Londonderry
BT 52 1SA Northern Ireland

new

SOIL TECHNOLOGY SERIES 1

U. Schwertmann, R.J. Rickson & K. Auerswald (Editors)

SOIL EROSION PROTECTION MEASURES IN EUROPE

Proceedings of the European Community Workshop on Soil Erosion Protection, Freising, F.R. Germany May 24 - 26, 1988

SOIL TECHNOLOGY SERIES 1

hardcover/224 pages/numerous figures, photos and tables

ISSN 0936-2568/ISBN 3-923381-16-6

list price: DM 119.-/US $ 75.-

ORDER FORM

☐ Please send me at the rate of DM 119.-/ US $ 75.-) .... copies of SOIL TECHNOLOGY SERIES 1.

☐ I want to enter a standing order for SOIL TECHNOLOGY SERIES starting with no. 1 (30% reduction on the list price)

Name ....

Address ....

Date ....

Signature: ....

Please charge my credit card: ☐ MasterCard/Eurocard/Access ☐ Visa ☐ Diners ☐ American Express

Card No.: .... Expiration date: ....

Please, send your orders to:

CATENA VERLAG, Brockenblick 8, D-3302 Cremlingen-Destedt, West Germany, tel. 05306-1530, fax 05306-1560

USA/Canada: **CATENA VERLAG**, Attn. Denize Johnson, P.O. Box 368, Lawrence, KS 66044, USA, Tel. (913) 843-1234, fax (913) 843-1244

THE GEOMORPHOLOGY
OF COASTAL SAND DUNES IN SCOTLAND
A REVIEW

M. Robertson-Rintoul & W. Ritchie, Aberdeen

Summary

There is abundant literature on Scottish sand dunes with most of it dating from the 1970's. Arguably there is more baseline, descriptive geomorphological data than for any other country in Europe. In contrast process-based research is uncommon. The exceptions are: studies of machair evolution in the Hebrides which are essentially paleoenvironmental; studies of dune ridge orientations and analysis of crest alignments and resultant wind vectors for selected dune systems in Scotland; and the study of airflow characteristics over parabolic dunes at Forvie where real-time airflow measurements were made and correlated with quantified sand movements.

Previous studies

Sand dune research in Scotland has been either ecological or a compilation of regional inventories.

It is primarily in the ecological literature that consideration is given to the primary development of dunes from the stage of sand accumulation above High Water Mark. This is the first trend distinguishable in Scottish dune literature. Embryonic dunes are equated with stands of *Agropyron junceiform*, mobile or yellow dunes with stands of *Ammophila arenaria*, while communities of *Elymo ammophiletum* are seral in nature and represent development from the yellow to fixed or grey dune stage (GIMINGHAM 1964). GIMINGHAM notes that research in Scotland has also drawn attention to geomorphological phenomena additional to the processes belonging to the primary succession. He refers in particular to the formation of erosional corridors or 'blow-outs' and their subsequent stabilisation.

Some ecological studies are distinguished by their consideration of the physical characteristics of the dune, the processes influencing dune morphology, and the history of the development of the dunes. These attempts to gain a general understanding of the geographical-historical-vegetational complex in individual areas is the **second trend** discernible in Scottish sand dune research. Nevertheless the total number of published surveys of this kind are few. Early in present century OGLIVIE (1914, 1923) and STEERS (1937) considered the Moray Firth area. The interrelated ecological and geographical problems presented by the dynamic nature of the Forvie Dune system has been discussed by LANDSBERG (1956). RITCHIE (1967, 1968, 1976, 1985) has examined the distinctive sand dune systems (Machair) in the Uists in detail.

In the 1970's, systematic inventories of **all** 647 beach and sand dune systems in Scotland were described in 13 regional reports (see bibliography) for the Countryside Commission for Scotland. MATHER & RITCHIE (1977 and 1984) used a morphological sequence in Scotland consisting of the parallel zones of backshore, dune, machair (dune pasture) and transition area as the basic descriptive mode (fig. 1). This sequence embodies the form of an idealised prograding dune coastline and contains the basic elements that RITCHIE (1972) suggests as being the initial topography of most of the Scottish dune systems. Nevertheless there are many deviations from this model, especially in the frequent truncation of the system where foredunes are absent — a form that is interpreted as a sign of coastline erosion. The variety of toporaphic settings e.g. bayhead, cliff foot, estuarine etc. also produces diversity of sand morphology.

In 1988 and 1989 five reports for the Nature Conservancy Council mark the beginning of research whereby the earlier reports are used as baselines to measure and to assess dynamic geomorphological change after more than a decade. The prime technique used is precise photomapping (HARRIS & WRIGHT 1987).

The dominant processes operating on the Scottish dunes today are erosional. From a break in the foredune vegetation there may follow a widening of the breach, slumping, linear extension and eventual removal of sand down to the base-level of the water-table or other sub-sand surface. In this way the simple parallel ridges of the Scottish dune systems are dissected by blowouts. Associated redepositional hillocks and ridges are found to leeward or behind the ridges. The blowouts may assume a variety of forms. Maximum relief diversity is produced by the resultant mosaic of flats and intervening ridges. RITCHIE (1972) suggests that it is this relatively chaotic mosaic which morphologically unites the Scottish dune systems.

A feature of special importance for coastline evolution in Scotland, including dunes, is the legacy of excess sedimentation both onshore and nearshore in the glacial and post-glacial periods. These factors have been described in full by many writers e.g. STEERS (1973), OGILVIE (1923), JARDINE (1964), CROFTS (1972) and RITCHIE (1966), and need not be discussed at length here. Nevertheless, it is important to emphasis that recent evidence of widespread erosion of foredunes may be linked to relatively recent sand deficiency. In short, energy, formerly used for massive post-glacial dune development at a time of unprecendented volumes of sand in coastal budgets (aided by rising sea levels), has been deflected to erosional processes which are most evident in the prevalence of retreating beach-dune interfaces and/or blowout development.

1 Process studies in Scotland

With few exceptions no process studies had been completed in Scotland before the 1980's. The exceptions were wind-tunnel modelling of sand movements which were field tested in real dune environments (PHILLIPPS & WILLETTS 1976). ESLER (1976) studied foredune stability at Forvie where he tested eight possible variables as being linked to sand movement. After a long series

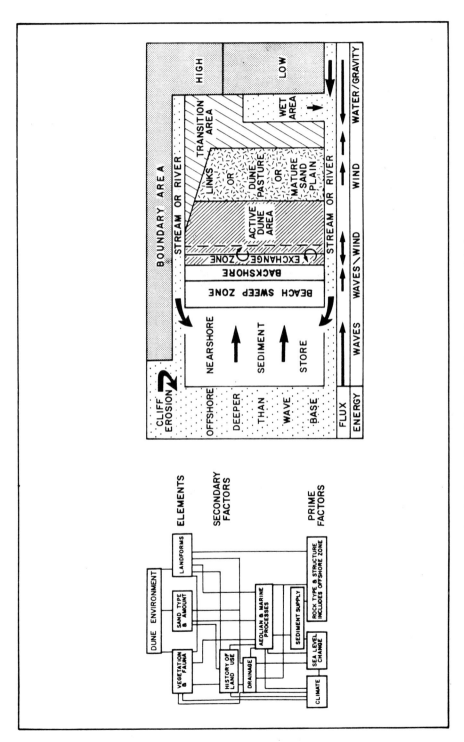

Fig. 1: *General model of sand dune elements and linkages.*

of repeated observations and measurements (but not process masurements) RITCHIE (1972) defined the essence of dune face erosion in vegetated dunes as the winnowing of loose sand from below the soil and root mat followed by slumping of the undercut edge. Loose dry sand accumulates in the vicinity of the erosional face. This loose sand may remain in place, may be moved alongshore, or may be redeposited on the upper dune edge by upwardly deflected airstreams. The latter will result in an increase in dune height, so that maximum relief in the dune system is associated with major blow-outs and active erosional processes in a variety of dune configurations.

Once initiated, erosion may proceed rapidly to base-level. This may be a shingle basement or a raised beach deposit. More often it is the water level. The end-product is the development of a low relief deflation plain.

The material deflated from the blowout is deposited downwind in a variety of forms. In the multidirectional wind regime characteristic of many of the Scottish systems, amorphous hillocks are common. In some areas, however, distinctive elongated redepositional forms, or whalebacks, are accumulated on the lee-slopes of the dunes.

The geometry of this suite of secondary forms does vary but the overall effect is one of an aeolian reshaping of the initial ridge. As such it is to be expected that both their orientation, pattern and distribution will reflect the vectors of wind energy.

The alignment of the fore dune ridge is determined by the alignment of the wave formed berm. Inland from the beach, once sand movement has been initiated the direction of sand drift will be the determining factor in the alignment of the resultant erosional and redepositional landforms. In general, sand movement will be either in a landwards or an alongshore direction. Three major factors combine to inhibit the large scale transference of sediment in a seawards direction in the vicinity of the foredune:

1. The influence of local shelter effects of the landward dune ridges, unless they are exceptionally low;

2. the frictional effects of vegetation;

3. the landward face of the foredune retains its vegetation cover, unless the blow-out has completely breached the ridge, or redeposition has been sufficiently rapid to result in slipface development.

Here the concept of effective onshore winds is applicable. Blow-out and whaleback orientation would be expected to reflect the resultant vector of onshore and alongshore winds. Results of the measurement of the alignment of the upper beach, blowouts and whalebacks together with the regional wind resultants for major dune systems has been completed for most major systems in Scotland by ROBERTSON-RINTOUL (1986). The preferred orientation of blow-out long axes and the crest lines of the whalebacks normally show a swing in direction along the length of most of the bays; this swing corresponds with the change in direction of the onshore wind resultant defined with respect to beach orientation. This close association between the direction of the resultant onshore wind and the alignment of the erosional and redepositional landforms indicates that these winds are exerting the controlling directional influence on the dune sand forms.

Coastal Sand Dunes in Scotland

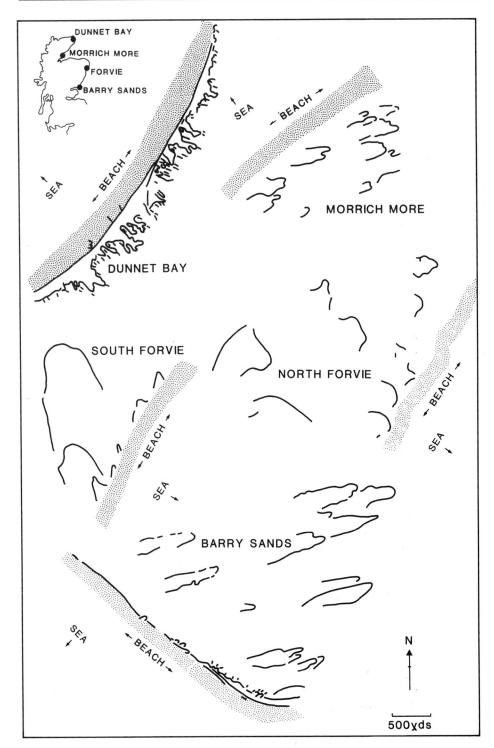

Fig. 2: *Planimetric maps of dune crestlines, east Scotland.*

The relation between dune orientation at the coastal edge and further inland for four sample areas is shown in fig. 2. This directional change normally involves a reorientation of dune forms to correspond with the vectors of onshore wind energy with increasing distance from the coastal edge.

These erosional and redepositional landforms represent a secondary or transgressive phase of dune ridge development. Dune systems characterised by the presence of these reshaped forms, should therefore be termed secondary dune systems. At this stage in the evolution of the dune sstem a net landward transfer of sediment is taking place.

2 Parabolic dunes in the sands of Forvie

More rarely in Scotland this net landward transfer of sediments is expressed in the development of dunes that are U or V-shaped. These dunes are or were partially vegetated with the spatial distribution of the vegetation being such that there is an increase in the percentage cover of vegetation outwards from the dune apex to the trailing arms. This results in an increase in surface roughness length from the dune centre to the arms. Wind velocity near the surface is reduced and so, therefore, are the sand transporting capabilities of the wind. The necessary prerequisite for sand movement, namely, bare sand in an unconsolidated form, is thus met primarily in the centre of the dune which migrates downwind relative to the arms giving rise to the characteristic U or V shape of parabolic dunes. The direction of this migration is indicated by the long axis of the dune.

The dune systems at the Sands of Forvie, Aberdeenshire (fig. 3), are dominated by U and rake-shaped dunes. The main sources of sediment supply for the dunes are the beaches at the mouth of the Ythan Estuary, located at the southern end of the Forvie Peninsula. The South Forvie dunes lie in relatively close proximity to the sand source. In contrast, the North Forvie parabolics are isolated from the sand source. Along the North Sea coastline of South Forvie a series of coastal edge blow-outs have cut through the coastal foredune. Behind the foredune there are two massive U-shaped dunes that occupy most of the area of the South Forvie Peninsula. The preferred orientation of the coastal edge blow-outs is 170° and the inland dunes 173°. The close correlation between the orientation of the blow-outs and U-dunes and the resultant onshore vector of 174° suggests that the onshore winds which blow over the source of sand play an important role in the formation of, and orientation of, the U-dunes and coastal-edge blow-outs of South Forvie.

Turning to North Forvie, the preferred orientation of the parabolic dunes is 247°. A clockwise swing in dune alignment from that characteristic of South Forvie is apparent. An explanation for this change in orientation may lie in a change in the direction of winds that are effective in dune shaping. In South Forvie, the dune forms are developed adjacent to the main source of sand, namely, the beach. However, North Forvie is isolated from the original sand source. The disposition of the sand source is no longer a crucial factor in defining the effective wind in dune formation. The effective winds can now be defined in terms of the duration and speed of the wind. If this hypothesis is correct then the preferred orientation of

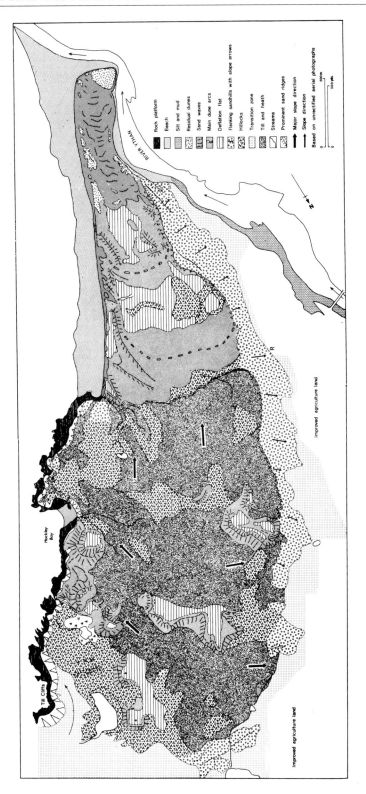

Fig. 3: *Generalised dune and coastal elements at Forvie.*

the North Forvie parabolic dunes should show a close correspondence to the resultant wind vector from all compass points. The preferred orientation of the parabolics and the resultant wind vector of 256° shows a fairly close correspondence.

The calculated effective wind vector resultants quoted above were derived using data from all the speed classes above the threshold wind speed for sand movement (11.6 knots). The sand rose concept as outlined by BAGNOLD (1951) and FRYBERGER (1979) thus has the limitation that the effect on dune shape of winds of varying strength cannot be evaluated. To gain some preliminary insight into this problem vector resultants from the Forvie wind data were calculated in two groups. The vector resultant for the speed classes below 22 knots was 248° and that for the speed classes above 22 knots was 265°. The lower wind speed classes account for 89 per cent of the percentage frequency of occurrence for winds above the threshold velocity. The higher speed classes account for only 11 percent. The very close correspondence of the preferred orientation of the parabolics to the resultant vector of lower magnitude winds suggest that these winds may have a predominant role in determining the orientation of the North Forvie parabolic dunes. The varying controls determining the alignment of the Forvie dunes suggest that the problem, noted in the literature, of the swing in dune orientation from coastal edge to inland parabolic dune, may be explained in terms of a change in the winds that are effective in dune shaping. At the coastal edge the influence of offshore winds is inhibited while the importance of the onshore winds is increased by virtue of their function in the transfer of the sediment from the upper beach nourishment zone of the dunes. As the dunes migrate inland they are isolated from the original sediment source by the sheltering effect of the dunes to seaward. The alignment of the dunes reflects to a greater extent the resultant of all wind vectors.

References

BAGNOLD, R.A. (1951): The physics of blown sand and desert dunes. 265 pp. Chapman and Hall.

CROFTS, R. (1972): Coastal processes and evolution around St. Cyrus. Unpublished M. Litt. thesis, University of Aberdeen.

ESLER, D. (1976): Coastal dune stability. Unpublished M.Sc. Dissertation, University of Aberdeen.

GIMINGHAM, C.H. (1951): Contributions to the maritime ecology of St. Cyrus, Kincardineshire. Part II. The sand dunes. TRans. Bot. Soc. Edinb. 35, 387–414.

GIMINGHAM, C.H. (1964): The maritime zone in the vegetation of Scotland. Ed. J.H. Burnett, Ch. 4, Edinburgh.

HARRIS, T. & WRIGHT, R. (1987): Change-detection in physically fragile coastal areas of Scotland by remote sensing. Proceedings of International Symposium on Remote Sensing of the Coastal Zone, Dept. of Geographic Information, Quensland, Australia. 16 pp.

OGLIVIE, A.H. (1914): The physical geography of the entrance to inverness firth. Scot. Geogr. Mag. 30, 21.

OGLIVIE, A.H. (1923): The physiography of the Moray Firth Coast. Trans. Roy. Soc. Edinburgh 53(2), 377–404.

PHILLIPS & WILLETTS (1976): Review of literature on sand stabilisation. Dept. of Engineering, University of Aberdeen. 19 pp.

RITCHIE. W. (1967): The Machair of South Uist. Scot. Geogr. Mag. 83(3), 162–173.

RITCHIE, W. (1968): The coastal morphology of North Uist. O'dell Mem. Mon. 1, 32 pp. University of Aberdeen.

RITCHIE, W. (1976): The meaning and definition of Machair. Trans. Bot. Soc. Edinburgh 42, 431–440.

RITCHIE, W. (1985): Scottish beaches and dunes. A national survey for recreational management purposes. Proc. 9th Cone Coastal Soc., New Jersey, 387–395.

RITCHIE, W. & MATHER, S. (1984): The beaches of Scotland. Countryside Commission for Scotland, 130 pp.

ROBERTSON-RINTOUL, M.J. (1985): The morphology and dynamics of parabolic dunes within the context of the coastal dune systems of Mainland Scotland. Unpublished D.Phil thesis, Oxford.

STEERS, A. (1937): The Culbin sands and Burghead Bay. Geogr. J., 498–528.

Addresses of authors:
M. Robertson-Rintoul
Environmental Consultancy Services Limited
52 Guild Street
Aberdeen AB1 2NB
Scotland UK
W. Ritchie
University of Aberdeen
Dept. of Geography
Elphinstone Road
Aberdeen AB9 2UF
Scotland UK

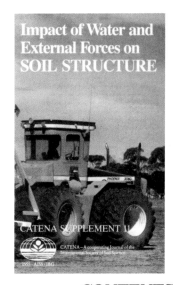

J. Drescher, R. Horn & M. de Boodt (Editors):

Impact of Water and External Forces on SOIL STRUCTURE

CATENA SUPPLEMENT 11, 1988

DM 149,— / US $ 88,—

ISSN 0722-0723 / ISBN 3-923381-11-5

CONTENTS

Preface

W.F. Van Impe, M. De Boodt & I. Meyus
Improving the Bearing Capacity of Top Soil Layers by Means of a Polymer Mixture Grout

H.H. Becher
Soil Erosion and Soil Structure

H.-G. Frede, B. Chen, K. Juraschek & C. Stoeck
Simulation of Gas Diffusion

H. Bohne & R. Lessing
Stability of Clay Aggregates as a Function of Water Regimes

A.R. Dexter
Strength of Soil Aggregates and of Aggregate Beds

R. Horn
Compressibility of Arable Land

K.H. Hartge
The Reference Base for Compaction State of Soils

B.G. Richards & E.L. Greacen
An Example of Numerical Modelling – Expansion of a Root Cavity in Soil

A. Ellies
Mechanical Consolidation in Volcanic Ash Soils

H.H. Becher & W. Martin
Selected Physical Properties of Three Soil Types as Affected by Land Use

I. Håkansson
A Method for Characterizing the State of Compactness of an Arable Soil

C. Sommer
Soil Compaction and Water Uptake of Plants

W. Köppel
Dynamic Impact on Soil Structure due to Traffic of Off-Road Vehicles

W.E. Larson, S.C. Gupta & J.L.B. Culley
Changes in Bulk Density and Pore Water Pressure during Soil Compression

A.L.M. van Wijk & J. Buitendijk
A Method to Predict Workability of Arable Soils and its Influence on Crop Yield

N. Burger, M. Lebert & R. Horn
Prediction of the Compressibility of Arable Land

H. Borchert
Effect of Wheeling with Heavy Machinery on Soil Physical Properties

P.H. Groenevelt
Impact of External Forces on Soil Structure

B.P. Warkentin
Summary of the Workshop

TRENDS IN THE DEVELOPMENT OF SAND DUNES ALONG THE SOUTHEASTERN MEDITERRANEAN COAST

H. Tsoar, Beer Sheva

Summary

Sand dunes of the southeastern Mediterranean coast are largely composed of sand brought to the sea by the Nile. The main factors which affect dune development are onshore winds and the size, inclination and continuity of the sea cliff.

The coastal dune strip is about 1000 years old. It is primarily affected by strong southwesterly and westerly winter winds. In this unidirectional wind regime, transverse and barchan dunes are the most common dune forms and are the first to be formed.

In summer, however, the dominant wind is an onshore sea breeze, orthogonal to the shoreline. Its effect is manifested on the southern horn of the barchans or transverse dunes, which is located so that the two dominant wind directions of winter and summer impinge upon it obliquely from both directions, thus elongating the horn into a seif. The coastal dunes along the Israeli shoreline are easily stabilized by *Acacia cyanophylla* except for the crest of the seif dunes.

ISSN 0722-0723
ISBN 3-923381-23-9
©1990 by CATENA VERLAG,
D-3302 Cremlingen-Destedt, W. Germany
3-923381-23-9/90/5011851/US$ 2.00 + 0.25

1 Introduction

The coastal dunes of the southeastern Mediterranean coast lie parallel to the shoreline, in a narrow strip up to 6 km wide, from the eastern part of the Bardawil Lagoon in northern Sinai up to Tel Aviv. This dune strip stretches between arid and semi-humid climates. The average annual rainfall along the northern Sinai coast is 90 mm, increasing to 600 mm along the northern coastal plain of Israel. All the rain falls between September and May. The shoreline in Israel is characterized in several places by a steep cliff of aeolianites. Encroachment of sand along the coast between Tel Aviv and Haifa is limited to breaches in the sea cliff (fig. 1).

The coastal sand dunes in Israel encompass a total area of 462 km^2, the Sinai coastal dunes cover a smaller area of 200 km^2.

Active coastal sand dunes throughout the world are known to be recent. It was only after 6500 to 3000 years BP that the sea reached its present level after the last rapid postglacial rise (GOLDSMITH 1985, ILLENBERGER & RUST 1988). In Israel and Sinai, the dunes of the coastal plain overlies soils containing numerous Hellenistic to Roman-Byzantine artifacts (ISSAR 1968, DOTAN 1982). It is also known that the coastal dunes of Caesarea (fig. 1) were used during the late Moslem leriod for primitive agriculture (PORATH 1975). This indicates that aeolian sand started to encroach on the coastal plain between the 7th and 9th centuries A.D., so that the maximum age of the southeastern Mediterranean coastal dunes is about 1000 years.

2 Factors affecting the rate of aeolian sand encroachment on the coastal plain

2.1 Vegetation

It has been taken for granted that encroachment of sand occurs during a dry spell, while cessation of sandflow onto the dunefield and stabilization of sand is indicative of a change in climatic conditions toward greater humidity (GERSON 1982, MAGARITZ 1986). However, the easy and deep penetration of rain in sand increases the leaching effect and reduces to a minimum the amount of essential plant nutrients such as nitrogen nd phosphorus. In addition, sand has, due to its coarse texture, only small quantities (3–9%) of available water for plants (NOY-MEIR 1973, BRADY 1974, 174). Therefore, there is little difference in the amount of moisture in sand in humid areas relatively to arid areas. Hence, vegetation on sand is meagre in all climates; therefore, there is no drastic difference in the amount of vegetation on dune sand in Israel between arid areas, with annual average rainfall of 150 mm, and humid areas with 600 mm. Encroachment of sand to form dunes can occur in humid areas with 2000 mm rainfall (HUNTER et al. 1983), as well as in the humid tropics (SWAN 1979).

Vegetation reduces the amount of sand transport. Measurements of the inland transportation of aeolian sand show that sand trapped by vegetated sand dunes is between 10% and 50% of the quantity trapped at the unvegetated beach/dune interface (GOLDSMITH et al. 1988).

One of the main factors hindering the growth of vegetation on dunes is moving sand; e.g. erosion by wind or accumulation of sand at a rate of >0.6 m yr^{-1}. Therefore, areas with high wind energy are generally devoid of vegetation even under humid conditions, as in the Oregon coastal sand dunes (HUNTER et al. 1983) and the Alexandria coastal dunefield in South Africa (ILLENBERGER & RUST 1988).

2.2 Source of sand

An important contribution to sand encroachment is sand availability. Rivers conveyy sand to the oceans and allow sand dunes to develop near their outlets according to the direction of longshore currents. The Nile is the main source of sand for the southeastern Mediterranean coastal dunes. This was demonstrated by POMERANCBLUM (1966), using heavy mineral analysis, and by EMERY & NEEV (1960), who showed a decrease in grain-size northwards to Tel Aviv and a uniformity of grain-sizes up to Haifa, which they attributed to the eroded aeolianite cliffs which serve an additional source. Nile sediments are carried eastwards by longshore currents whose north-eastward component gradually decreases in Israel proportionately to the distance from the Nile (GOLDSMITH & GOLIK 1980).

Deposition of sand in the Haifa bay (between Haifa and Akko, see fig. 1) is due to a longshore sand "sink" with

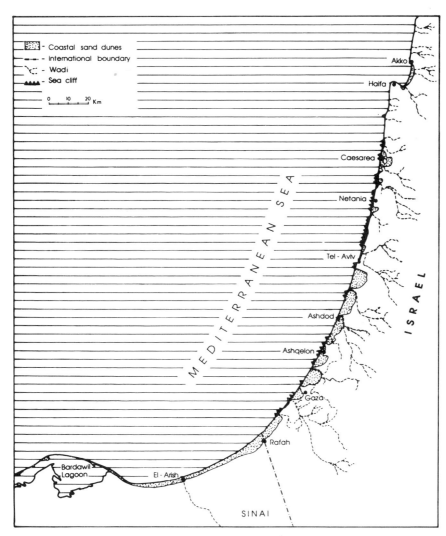

Fig. 1: *Coastal sand dunes of Israel and Sinai. Note the wider coastal dune strip in Israel and the masses of sand south of Tel Aviv which are blocked by wadis at their northern side.*

large quantities of sand trapped by wave refraction (GOLDSMITH & GOLIK 1980). According to Bronze age sites (KOLNER & OLAMI 1980), the Haifa beach has widened by about 4 km since the Middle Bronze age (3500-4000 B.P.) causing the growth of a large dune system.

2.3 Wind direction and magnitude

The wind magnitude and duration are very important factors affecting the encroachment of sand. For coastal sand dunes to develop, strong onshore winds are required. Measurments taken recently along the beach/dune interface in

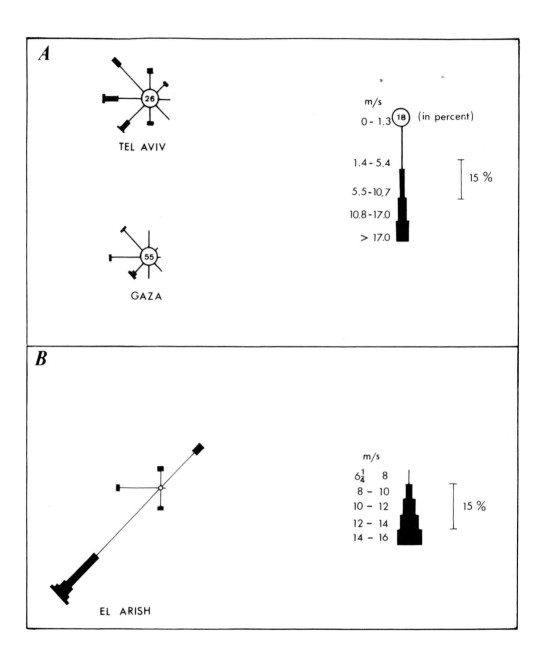

Fig. 2: *(A) Annual wind roses for Tel Aviv and Gaza for 0800, 1400 and 2000 hours (after ATLAS OF ISRAEL 1956). (B) Annual wind rose for magnitudes above 6.25 m s^{-1} (threshold velocity), recorded in El-Arish, northern Sinai (after TSOAR 1974).*

Israel reveal that the inland transport of aeolian sand transport (having moisture of 1.05% by weight) starts at a wind velocity of 10 m s^{-1} measured at a height of 2 metres. Under falling rain, the sand moisture increased to 2% and there was no sand movement at all, even at wind velocities of 25 m s^{-1}. Substantial aeolian sand movement from the beach inland only occurs under the strong winds.

A demonstration of the effect of wind on the formation of coastal dunes is given in fig. 1. Since it is known that the source of the sand is the Nile, the Sinai coastal dune strip should be wider than more distant dunes strips in Israel. However, as shown in fig. 1, the dune strip most distant from the Nile is wider (5–6 km) than those in Sinai (2–3 km). Obviously, the wind regime prevailing in that area accounts for this phenomenon (fig. 2). The strongest winds occur in winter and blow from SW and S directions. In Sinai the shoreline runs parallel to the direction of the winter storms, although further north and east, in Israel, the beach becomes orthogonal. The summer sea breeze with low velocities is orthogonal to all coastlines, independent of geograghical orientation. Recent field measurements on the coastline of Israel between Gaza and Akko (175 km long of which 104 km contain dunes) indicate that the net transport of aeolian sand from beach to dunes is 42,000 m^3 yr^{-1} (GOLDSMITH et al. 1988).

2.4 Sea cliff

Some of the former Pleistocene cycles of aeolian sand encroachment onto the coastal plain of Israel have ended in consolidation of the dunes into aeolianites which compose the dominant rock along the shoreline. Aeolianites were formed in areas having annual average rainfall from less than 300 mm to just over 600 mm (YAALON 1967). For this reason they are found in Israel and not in Sinai. Subsequent sea level changes and wave erosion has produced a cliff situated at the seaward edge of the coast. North of Tel Aviv, the sea cliff is high (up to 50 m) and continuous, and prevents inland encroachment of beach sand by wind. Only in areas where this natural impediment is either nonexistent or breached by streams, is sand allowed to penetrate and coastal dunes develop (fig. 1).

South of Tel Aviv, sea cliffs are less common and, where they occur, are low and discontinuous. The coastal dunes in this area form seven units which are limited at their northern boundary by a wadi. The one exception is in the area north of Ashqelon, where the northern limit of the dune is determined by the existence of sea cliffs on the beach which permit only minor encroachment of sand (fig. 1).

Measurements of aeolian sand transport were made during strong winter winds of 25 m s^{-1} (measured at height of 2 m) using traps. It was shown that sand is incapable of climbing a cliff with an inclination of 30° to 40° and a height of 22–25 metres. The wind impinges on the cliff at angles of 45° to 50° to the cliff edge and is diverted to flow parallel to the shoreline along the foot of the cliff. The cliffs south of Tel Aviv are discontinuous and usually less than 300 m long. The diverted sand-moving wind is deflected again at the northern end of the cliff, which causes it to go around the cliff and flow in an inland direction. In this way the aeolian sand eventually encroaches inland.

Summarizing, it is concluded that

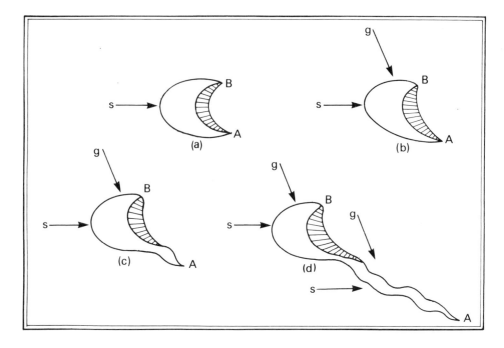

Fig. 3: *Development of an elongated horn of a barchan that eventually turns into a seif dune through the action of a secondary gentle wind (after TSOAR 1984).*

the formation of coastal dunes along the southeastern Mediterranean coasts is subject, on one hand, to the delivery of adequate supplies of sand by longshore currents and, on the other hand, to storm winds blowing orthogonal to the shoreline and being capable of moving sufficient sand. Steep continuous cliffs prevent inland encroachment of aeolian sand. When the sea cliff is discontinuous, sand penetrates and coastel dunes form.

3 Coastal dune types and trends in their development

The accretion of sand into sand dunes is only caused by strong winds (BAGNOLD 1941). On coasts, as in Israel and Sinai, where strong winds from only one direction prevail, the initiation and formation of coastal dunes obey such unidirectional orientation, resulting in transverse dunes or, if covered by vegetation (as occurs mostly near the shoreline), in foredunes and parabolic dunes.

The coastal transverse dunes in Israel are orientated according to the SW and W storm winds. Once built up, they are also affected by the summer sea breeze, a more gentle wind as it occurs in Israel and Sinai, blowing always on dry sand at angles between 45° and 120° to the strongest SW and W winter winds (figs. 2 and 3). Dunes are thus subject to a bidirectional wind regime. A model for the development of transverse and barchan dunes based on these circumstances is shown in fig. 3a and b, starting with its initiation and formation by the stronger SW winter winds (s). Horn B is eroded

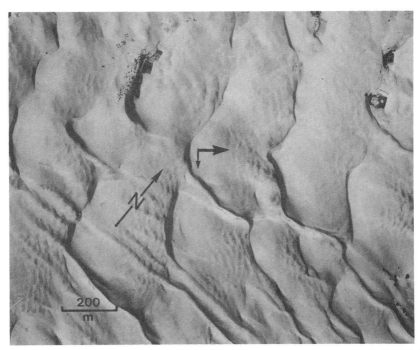

Photo 1: *An aerial photograph of a coastal crescent-shaped transverse dunefield in northern Sinai (average annual rainfall is 175 mm).*

The elongated southern horns of the transverse dunes represent the origin and initiation of seif dunes formed from transverse dunes as a result of a bi-directional wind regime. The thick arrow indicates direction of the SW winter storm winds that formed the transverse dunes initially, wheras the thin arrow indicates the direction of the NW summer sea breeze. The formation and elongation of seif dunes respond to these two directions.

by gentle sea breeze coming from direction g. Horn A is located on the side opposite to these gentle winds, so that two wind directions (s and g), encounter it on either side.

This type of correspondence between sand dune and the impact of sand-moving wind is similar to how seif dunes develop (TSOAR 1983). The two wind directions (s and g) develop longitudinal movement of sand along both lee sides of horn A (fig. 3b and c), elongating it to form a seif dune (fig. 3d).

When seif dunes are formed from barchan or transverse dunes, the result is two aeolian bedforms on one dune system, each of which has its particular mechanism of movement. Transverse and barchan dunes advance by means of erosion of sand on the windward slope and its deposition on the lee slope (slip-face). Seif dunes do not advance in this manner, but by elongation of the previous dune form. From a dynamic viewpoint, seif dunes have an advantage over transverse or barchan dunes in that any wind direction is deflected on the lee side to move on parallel to the crest line, thus bringing about the seif's elongation.

Consequent to the above process, seif

Photo 2: *A seif dune in the southern part of the Gaza Strip. Note the sharp crest which is the most active part of the dune.*

dunes elongate faster than the advance rate of connected sections of transverse or barchan dunes (photo 1). The young age (about 1000 years) of the coastal dunes of Israel and Sinai may account for the dominance of transverse dunes there. It takes probably several hundred years for a transverse dune to fully develop a long seif dune on its horn.

The process of development of elongated horns of transverse dunes into seifs is a more frequent phenomenon in the Sinai coastal dunes (photo 1) than in the Israeli coastal dunes. This is due to the different direction of the sea breeze in Sinai relative to that in Israel caused by the crescentic shoreline. In Sinai the two dominant wind directions, those of winter and summer, encounter each of the two slopes of the initiated seif dune at angles closer to the critical ones needed to cause maximal longitudinal movement of sand on the lee slope (TSOAR 1978) and thus increase the seif's rate of elongation. The lack of vegetation on the coastal dunes of Sinai, relative to Israel, also encourage the formation of seifs in Sinai.

4 The effect of the shape of the dune on the generation of vegetation

The dominant dune types in the coastal plain of the southeastern Mediterranean — the transverse and seif dunes — are formed on sand devoid of vegetation. They preserve their typical morphology as long as they are not covered by vegetation (photo 1).

Vegetation on sand is very vulnerable to erosion by wind, which denudes the roots; a process which generally kills the plant. Rain water on sand penetrates to a great depth and moisture is thus protected from evaporation during the long dry seasons. For that reason, as long as the wind magnitude is moderate, sand in arid and semi-arid areas allows for relatively more vigorous and more perennial mesophytic shrubs than fine-textured soils. Hence, erosion of sand, and not lack of moisture, is the primary limiting factor for vegetation on dunes, both in humid and in arid areas.

The rate of sand erosion depends on wind magnitude and dune shape. On a transverse dune the windward slope is bare of vegetation because of exposure to continuous erosion. Its rate of erosion is proportional to the sine of the slope's angle of declination (ALLEN 1968). This also explains why at the crest of the transverse dune, where the angle of declination is zero or near zero, there is neither erosion nor deposition of sand. Therefore, in the semi-arid areas of the coastal dunes in Israel, vegetation is only found on the crest of transverse dunes. On a seif dune (photo 2) the crest is its most active part; erosion there is almost continuous on both slopes and impedes the generation of vegetation.

The coastal sand dunes of Israel have suffered vegetation destruction for many years. Since 1921, various attempts have been made to stabilize them (TEAR 1925, 1927, WEITZ 1932). The *Acacia cyanophylla* is known to be a good sand stabilizer species for the semi-arid and sub-humid dunes of Israel. However, it is difficult for the *Acacia cyanophylla* to prosper on the crestal area of seifs which is most exposed to erosion.

Acknowledgement

I would like to thank Jan A. Klijn and Victor Goldsmith for helpful discussion and comments on the manuscript.

References

ALLEN, J.R.L. (1968): Current ripples. North Holland, Amsterdam.

ATLAS OF ISRAEL (1956): Survey of Israel, Tel Aviv.

BAGNOLD, R.A. (1941): The physics of blown sand and desert dunes. Methuen, London.

BRADY, N.C. (1974): The nature and properties if soils. 8th edn. MacMillan, New York.

DOTAN, T. (1982): Lost outpost of the Egyptian Empire. National Geographic **62**, 739–769.

EMERY, K.O. & NEEV, D. (1960): Mediterranean beaches of Israel. Bulletin Geological Survey of Israel **26**, 1–24.

GERSON, R. (1982): The Middle East: landforms of a planetary desert through environmental changes. Striae **17**, 52–78.

GOLDSMITH, V. (1985): Coastal dunes. In: R.A. Davis Jr.Coastal sedimentary environments. 303–378. Springer-Verlag, New York.

GOLDSMITH, V. & GOLIK, A. (1980): Sediment transport model of the southeastern Mediterranean coast. Marine Geology **37**, 147–175.

GOLDSMITH, V., ROSEN, P. & GERTNER, Y. (1988): Eolian sediment transport on the Israeli coast. Final Report, U.S.-Israel BSF, National Oceanographic Institute, Haifa, Israel.

HUNTER, R.E., RICHMOND, B.M. & ALPHA, T.R. (1983): Storm-controlled oblique dunes of the Oregon coast. Bulletin Geolgoical Society of America **94**, 1450–1465.

ILLENBERGER, W.K. & RUST, I.C. (1988): A sand budget for the Alexandria coastal dunefield, South Africa. Sedimentology **35**, 513–521.

ISSAR, A.S. (1968): Geology of the central coastal plain of Israel. Israel Journal of Earth Sciences **17**, 16–29.

KLONER, A. & OLAMI, Y. (1980): The early and middle Canaanite periods. In: A. Soffer and B. Kipnis, Atlas of Haifa and Mount Carmel. 34–35. Applied Scientific Research CO., University of Haifa.

MAGARITZ, M. (1986): Environmental changes recorded in the upper Pleistocene along the desert boundary, southern Israel. Palaeogeography, Palaeiclimatology, Palaeoecology **53**, 213–229.

NOY-MEIR, I. (1973): Desert ecosystems: Environment and producers. Annual Review of Ecology and Systematics **4**, 25–51.

POMERANCBLUM, M. (1966): The distribution of heavy minerals and their hydraulic equivalent in sediments of the Mediterranean continental shelf of Israel. Journal of Sedimentary Petrology **36**, 162–179.

PORATH, Y. (1975): Keyseri gardens. Qadmoniot **8**, 90–93 (in Hebrew).

SWAN, B. (1979): Sand dunes in humid tropics: Sri Lanka. Zeitschrift für Geomorphologie N.F. **23**, 152–171.

TEAR, F.J. (1925): Sand dune reclamation in Palestine. Empire Forestry Journal **4**, 24–38.

TEAR, F.J. (1927): Sand dune reclamation in Palestine. Empire Forestry Journal **6**, 85–93.

TSOAR, H. (1974): Desert dunes morphology and dynamics, El Arish (Northern Sinai). Zeitschrift für Geomorphologie N.F. Suppl. **20**, 41–61.

TSOAR, H. (1978): The dynamics of longitudinal dunes. Final technical report. European Research Office, U.S. Army, London.

TSOAR, H. (1983): Dynamic processes acting on a longitudinal (seif) dune. Sedimentology **30**, 567–578.

TSOAR, H. (1984): The formation of seif dunes from barchans — a discussion. Zeitschrift für Geomorphologie N.F. **28**, 99–103.

WEITZ, J. (1932): La fixation des dunes en Palestine. Silva Mediterranea **7**, 1–26.

YAALON, D.H. (1967): Factors affecting the lithification of eolianite and interpretation of its environmental significance in the coastal plain of Israel. Journal of Sedimentary Petrology **36**, 1189–1199.

Address of author:
Haim Tsoar
Department of Geography
Ben-Gurion University of the Negev
Beer Sheva
Israel

COASTAL DUNES IN DENMARK. CHRONOLOGY IN RELATION TO SEA LEVEL

Ch. Christiansen, K. Dalsgaard, J.T. Møller, Aarhus
D. Bowman, Beer Sheva

Summary

Coastal dune building in Denmark has been episodic. The Tapes/Littorina (flandrian) transgression apparently caused no dund building activity. On the contrary, the 3 dune building periods in Danish coastal areas were all associated with low sea-level. This suggests that low sea-level and availability of sediments play a more deterministic role in coastal dune building in Denmark than any anthropogenic effects. Today, with a mean sea-level rise of up to 1.3 mm/y, dune building takes place where sedimentation from the longshore sediment transport causes coastal progradation. An interesting example of dune building, on top of a 65 m high coastal cliff, is described.

1 Introduction

Coastal dunes occupy about 800 km² or about 3% of the total area of Jutland (fig. 1) demonstrating the role of eolian deposition in Danish coastal areas.

The formation of coastal dunes has three major requirements: a substantial sand supply, a strong onshore wind for transport for at least part of the year and an area in which sand can accumulate. It is evident from previous studies that dune formation in Denmark has been episodic (BRUEL 1918, CHRISTIANSEN & BOWMAN 1986), with periods of dune stabilization, weathering and pedogenese (BOWMAN, CHRISTIANSEN & MARGARITZ 1989). As yet, however, the precise nature and relative importance of the causative factors of dune formation have not been established with certainty.

The coastal dunes attained their present morphology more than 200 years ago and perhaps therefore have attracted only little attention. The coastal erosion which has taken place since then (BIRD 1974) has apparently never triggered major dune development. Thus, only few Danish papers deal with dune development in itself and most of the present knowledge on dune building periods comes from historical and archaeological literature.

2 Dune building periods

Old coastal dunes (>500 BC) are not found in Denmark. This might be due to the relative young age of the present coastal environment. Because of the isostatic rebound following the Ice Age, marine deposits from 11000–12000 B.C., have been raised up to 56 m in the north-

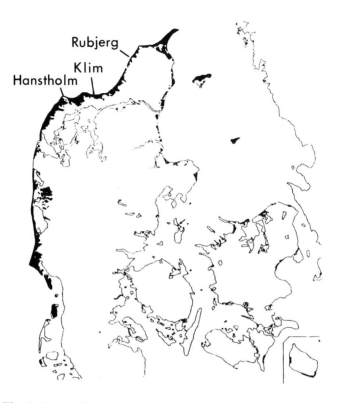

Fig. 1: *Map of Denmark showing distribution of coastal dunes.*

ern part of Denmark.

Along coastal areas in Western Denmark it is possible to identify at least three dune building periods. Near the base in many Danish coastal dunes a layer of peat is often present. This layer has been dated at 400 B.C. as well from pollen analysis as from the archaeological content (JESSEN 1899, JENSEN 1981). These datings were confirmed by C14 datings of 355–240 B.C. from the top of the layer (LIVERSAGE, MUNRO, COURTY & NØRNBERG 1987). The peat layer becomes gradually more sandy upwards and is covered by dune sand. This dune building episode coincides with a relatively cold period (AABY 1975) and a low sea-level (fig. 3).

The second dune building episode took place about 400–600 A.D., again in a period of relatively low sea-level (fig. 3). In the Hanstholm area (fig. 1) BOWMAN, CHRISTIANSEN & MARGARITZ (1989) dated the top of podzols, which were buried by dunes, at 449 ± 109 A.D. and 394 ± 128 A.D. An indirect evidence of the same dune building episode is found in the Klim area. Pits in a meadow, fenced in by limestone hills to the N, S and E, show a general statigraphy of 70 cm windblown sand on top of 90 cm of peat. The peat overlays a raised marine sandflat. Peat formation started at 685–760 A.D. This could indicate that somewhat before the end of the 7th century dune formation took place north of the present meadow, which stopped the only posible drainage to the north and thereby created the bog.

The top of the peatlayer in the Klim area became covered with windblown sand in 1645 A.D. In the Hansholm area podzols developed in the dune sand from about 400 A.D. and became covered with new dunes in 1486 A.D.

3 Dune formation

This last major period of dune formation in Denmark took place between 1450 and 1750. This coincides with the climatically distinct Little Ice Age. As the period is well documented historically, it might throw light on the relative importance of the causative factors. It is evident from contemporary literature (PONTOPPIDAN 1769) that the main cause was believed to be interventions in the biological system. Already in 1539 King Christian III passed a law which forbade farmers the use of dune vegetation. According to PONTOPPIDAN (1769), eolian activity took place because the farmers cut the dune vegetation to feed their animals, for use as fuel or for use as roof-cover. From the study of historical documents covering the period 1550–1750 BRÜEL (1918) concluded that the eolian activity was caused by overgrazing in the dunes and especially by forest cutting. HANSEN (1957) rejected the forest cutting theory as the regions which experienced most eolian activity were never forested. He agreed in that the cause was ruthless exploitation of dune veetation but found that the main reason was climatic worsening. This had forced the farmers to exploit the dune vegetation. The theory of forest cutting as a principal cause has also been rejected by several authors in The Netherlands (KLIJN 1981). A major shortcoming in these theories is that they fail to explain the main source of the huge amount of sand needed to cover 25% of the former Thisted county or to build dunes up to 16 km from the sea. The soils in this part of Denmark are formed on clayey moraines.

4 Models of dune development

Fig. 3 points to relations between sea-level and dune building processes. Previous work (PYE 1983) has suggested two main models for coastal dune development dependent on sea-level variation. The models are diagrammatized in fig. 2.

Model 1 suggests that major dune building takes place in periods with rising or high sea-level. COOPER (1958) found that rising sea-levels lead to beach erosion, the reworking of former dunes, the destruction of vegetation on dunes and beach ridges, the creation of blow-outs and result in transgressive dunes. Landward migration of barrier islands, triggered by wave erosion, overwash and wind drift might also result in transgressive dune formation (JELGERSMA & VAN REGTEREN ALTENA 1969, SANDERS & KUMAR 1975). Acording to this model, low sea-level is associated with dune stability, weathering and pedogenesis (BRETZ 1960).

Model 2 proposed by among others SCHOFIELD (1975), CHRISTIANSEN & BOWMAN (1986) and MARTIN (1988) suggests the opposite interpretation for sealevel - dune building relationships. According to this model major dune building takes place in periods of falling or low sea-level. The lowering of wave base makes more sand available for landwart transport, expose wide areas for eolian activity and results in

MODEL 1
RISING AND HIGH SEA LEVEL MODEL

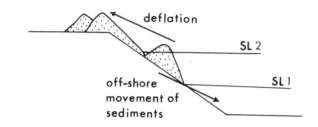

MODEL 2
LOW AND FALLING SEA LEVEL MODEL

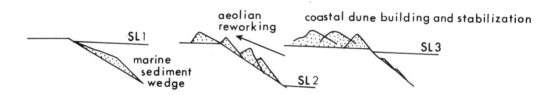

Fig. 2: *Two models of coastal dune formation related to sea-level change (after CHRISTIANSEN & BOWMAN 1986).*

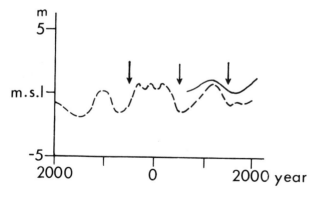

Fig. 3: *Sea-level variations in the south-eastern part of Denmark during the last 3 thousand years (after JACOBSEN 1987) and probable co-tidal line of HWNT (High Water Neap Tide) in northern Germany from 650 A.D. (after ROHDE 1978). Dune building periods are marked with arrows.*

shoreline progradation. During rising or high se-level the dunes are deprived of their sand supply and become stabilized.

The Danish dune building periods apparently fit into model 2. Fig. 3 shows a juxtaposition of sea-level variation and dune building periods during the last 4000 years. There is a clear coincidence between periods of relative low sea-level and periods of dune building. Similarly, relative high sea-level coincide with periods of peat formation and dune stabilization. Although the Danish transgressions do not fit in time with Dutch observations (JELGERSMA & VAN REGTEREN ALTENA 1969) they apparently in both countries fit with standstill phases of dune formation. Even with an allowance of 100-200 years in which dunes must have migrated from the coast to the locations where datings have been established, as suggested by HANSEN (1957), the dune building process must have taken place during falling sealevel (see fig. 3). MARTIN (1988) also noted that dune building in Estonia was related to stillstand of the shoreline or even to minor regressions during the Late-Holocene marine transgression of the Baltic. One relatively unknown factor here is the frequency of storms and stormfloods which may govern sand budgets even more than absolute sealevel.

During the 18th century further dune building was stopped by local and state authorities in many places by the establishment of plantations in dune areas. However, BRÜEL (1918) quotes several examples in which the sand drift stopped by itself in the late 18th century. This implies that the sea-level rise which started in the dune building period did not trigger model 1 dune building. On the contrary, the dunes become stabilized partly because wide previously subaerial areas become again covered by the sea.

5 Present situation

Sea-level is rising by up to 1.3 mm/y in Denmark (CHRISTIANSEN, MØLLER & NIELSEN 1985). This has triggered no major dune activity. A reason for this might be that in most places the dunes are carefully planted at any signs of blow-out tendency. However, also in places where there are no dune protection, there are no signs of dunes transgressing inland. Fig. 4 shows a beach profile 1968–1988 at Klim. (For location see profile 600 in CHRISTIANSEN & MØLLER 1980). It can be seen that the outer dune disappeared between 1981 and 1982 (during a storm surge in November 1981) and a new dune started to build up further 30 m inland. Closer inspection of fig. 4 shows that the dune transgression is only apparent, following beach erosion the beach and dune environment act as one system.

CHRISTIANSEN & BOWMAN (1990) examined coastal changes 1944–1987 along a 80 km stretch of coast in NW Jutland. Fig. 5 is based on their data. It can be seen that there is a strong correlation (r=0.93) between changes in water-line and dune front position. Where the water-line is eroded from 1944-87 there is an accompanying erosion of the dune front. On the other hand, areas which have a positive sediment supply to the beach show advancing windward growing dune fronts. Fig. 5 thus shows the importance of a large sediment supply in coastal dune building. As sea-level is rising fig. 5 also indicates that it is perhaps not sea-level variation directly in itself which causes dune formation. The effect of sea-level variation is apparently indirect, operat-

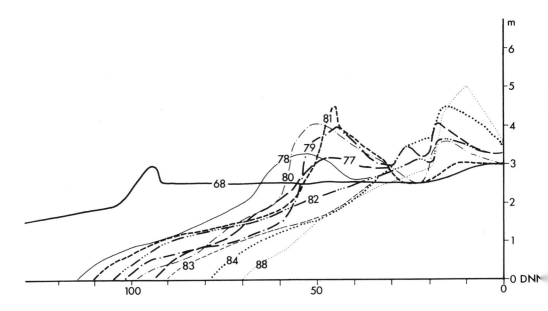

Fig. 4: *Changes in beach profile 1968–88 at Klim.*

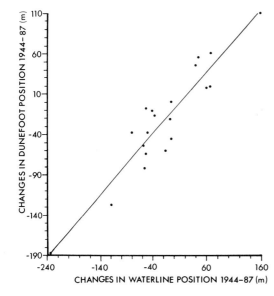

Fig. 5: *Regression of dunefoot changes 1944–87 on waterline (high water level) changes in the same period (r=0.93).*

Coastal Dunes in Denmark

Photo 1: *The highest part of the dune and the lighthouse seen from the south: the lighthouse is no longer visible from the sea due to dune building.*

Photo 2: *Rubjerg Knude seen from the east with the lighthouse in the far background and the cultivated field in the foreground. The burial mound is seen to the left situated on a low hill.*

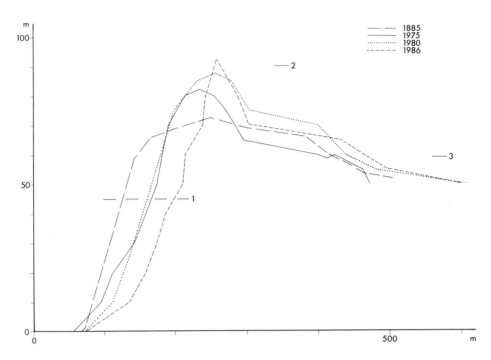

Fig. 6: *Cross sections of Rubjerg Knude from the years 1885, 1975, 1980 and 1986. 1: the level at which Viking Age tombs from 900 A.D. were excavated; 2: the beam height of the lighthouse; 3: the top of the burial mound.*

6 A case study of Rubjerg Knude

Rubjerg Knude is a rather remarkable dune situated in northern Jutland (photo 1). Contrary to other not completely vegetated dunes, Rubjerg Knude remains rather stable in position owing to its height. Originally the highest point on the coast was a moraine cliff consisting of sand and clay folded up by ice pressure to a height of 65 m above sea level. blown sand, coming from offshore bars, the beach and the cliff itself is moving up the steep cliff, which is frequently eroded by waves. The sand has been deposited on the clif top now situated at a height of 92 m above sea level. In 1900 a lighthouse was constructed near the cliff top of that time. In 1969 it had to be abandoned because the dune had grown so much in height that the light could not be seen from the North Sea.

Cross sections (fig. 6) of Rubjerg Knude near the lighthouse clearly show the rather extraordinary development. The dune has increased in height, but not much in extent, towards the east. From 1885 to 1975 the average growth in height of Rubjerg Knude was only 10 cm/yr, most likely with a very slow growth in the beginning of the period, accelerating later on. During 1975–80 the growth in height was more than 1 m/yr.

After 1980 the growth decreases probably because the dune is now so high in proportion to the short east-west extent that sand slides prevent a further net growth in height. From the cross sections, it can be seen that the development of the dune on the cliff top had not started in 1885 or was in an initial stage.

From the geological point of view, the time elapsed since the first reliable topographic maps surveyed in 1885, is very short. Fortunately, the development can be dated far back in history. Just east of the lighthouse is a cultivated field on very sandy soil. If the field has been subjected to strong erosion or sand accumulation it would have been abandonned long ago. Further, a burial mound is situated on a low hill on the southern edge of the field (photo 2). the tomb has never been investigated or excavated, but in Denmark this type of burial mound is very old, not later than approximately 500 A.D. and it can be hundreds of years older.

As mentioned above the moraine cliff top is exposed from time to time by erosion and sand slides. Some years ago, a narrow strip of the old moraine surface remained uncovered of sand for a short period. About one kilometre south of the lighthouse some ancient artifacts were found. During excavations for further finds, some tombs dated to about 900 A.D. were found in this surface. The bodies had been buried about 900 A.D., which is the Viking Age. These archaeological findings thus corroborate the idea that the area east of the present lighthouse has remained almost untouched by sand accumulation and erosion up to about 1885.

The explanation for the position of the dune is still not known, but most likely the height of the moraine cliff is the important factor. North and south of Rubjerg Knude the moraine cliff top is situated about 30 m above sea level. thus the original difference in height between the distal and the central parts of the moraine cliff is about 35 m. This difference could be sufficient to explain the change in sand transport up the cliff and the depositional conditions on the top.

Apparently very large desert dunes are rather stable because they affect the wind climate. Wind coming from one side of the dune creates vortices on the opposite side of the obstacle, and consequently winds mostly blow towards the summit. Possibly the steep cliff at Rubjerg Knude act as a large desert dune. Probably the height, at which vortice are strong enough to create winds in an opposite direction of the mainwind, can be found between 30 and 65 m on steep slopes in this environment.

A special feature of this cliff is important for the wind climate. As mentioned above the cliff is formed by alternating layers of sand and clay folded up by ice pressure. The clay deposits are rather stable, but the sand is easily removed by the wind, water erosion and slides. Therefore, the cliff has furrows acting as chimneys facilitating the formation of jets upwards the steep cliff. Anyhow, during strong winds from west sand can be observed moving upwards the cliff as well from west as well as from east. Much sand is sliding downwards again, but owing to the sparse vegetation on top of the dune, new sand deposits can be seen on the top after all strong winds.

References

AABY, B. (1975): Cykliske klimavariationer de sidste 7500 år påvist ved undersøgelser af højmoser og marine transgressionsfaser. Danmarks Geologiske Undersøgelser. Årbog 1974, 91–104.

BIRD, E.C.F. (1974): Coastal changes in Denmark during the past two centuries. Skrifter i Fysisk Geografi Nr. 8. University of Aarhus. 21 pp.

BOWMAN, D., CHRISTIANSEN, C. & MARGARITZ, M. (1989): Late-Holocene coastal evolution in the Hanstholm-Hjardemaal region, NW Denmark. Morphology, sediments and dating. Geografisk Tidsskrift **89**, 49–57.

BRETZ, J.H. (1960): Bermuda, a partially drowned, late mature, Pleistocene karst. Bulletin of the Geological Society of America **71**, 1729–1754.

BRÜEL, J. (1918): Klitterne i Vestjylland og på Bornholm. Nordisk Forlag, København. 133 pp.

CHRISTIANSEN, C. & BOWMAN, D. (1986): Sea-level changes, coastal dune building and sand drift, North-Western Jutland, Denmark. Geografisk Tidsskrift **86**, 28–31.

CHRISTIANSEN, C. & BOWMAN, D. (1990): Long-term beach and shoreface changes, NW Jutland, Denmark: Effects of a change in wind direction. In: Beukema, J.J. & Wolff, W.J. (eds.), Proceedings Workshop Climatic Effects on Coastal Ecosystemes. Balkema, Rotterdam. (In press).

CHRISTIANSEN, C. & MØLLER, J.T. (1980): Beach erosion at Klim, Denmark. A ten year record. Coastal Engineering **3**, 283–296.

CHRISTIANSEN, C., MØLLER, J.T. & NIELSEN, J. (1985): Sea-level fluctuations and associated morphological changes: Examples from Denmark. Eiszeitalter und Gegenwart **35**, 89–108.

COOPER, W.S. (1978): Coastal sanddunes of Oregon and Washington. Memoires of the Geological Society of America **72**, 169 pp.

HANSEN, V. (1957): Sandflugten i Thy of dens indflydelse på kulturlandskabet. Geografisk Tidsskrift **56**, 69–92.

JELGERSMA, S. & VAN REGETEREN ALTNA, J.F. (1969): Geological history of the coastal dunes in the W. Netherlands. Geologie en Mijubouw **48**, 335–342.

JENSEN, J.A. (1981): Træk af landskabets og den ældste bosætningsudvikling. In: Rasmussen, A.H. (ed.), Holmslands og klittens historie. II. Holmsland Kommune. 9–23.

JESSEN, A. (1899): Kortbladene Skagen, Hirtshals, Frederikshaven, Hjørring og Løkken. Geological Survey of Denmark. I Rk, Nr. 3.

KLIJN, J.A. (1981): Nederlandse kustduinen; geomorfologie en bodens. Thesis, Wageningen, 189 pp.

LIVERSAGE, D., MUNRO, COURTY, M.-A. & NØRNBERG, P. (1987): Studies of a buried early Iron Age Field. Acta Archaeologica **56**, 55–84.

MARTIN. E. (1988): relationship between dune formation and Baltic Sea transgression in Estonia. In: Winterhalter, B. (ed.), the Baltic Sea. Geological Survey of Finland, Special Paper **6**, 79–85.

PONTOPPIDAN, E. (1769): Den danske atlas eller kongeriget Danmark, Tomus 5, København.

PYE, K. (1984): Models of transgressive coastal dune building episodes and their relationship to Quaternary sea level changes: a discussion with reference to evidence from eastern Australia. In: Clark, M.W. (ed.), Coastal Research U.K. perspectives. Geobooks, Norwich. 81–104.

ROHDE, H. (1978): The history of the German coastal area. Die Küste **32**, 6–29.

SANDERS, J.E. & KUMAR, N. (1975): Evidence of shoreface retreat and in place 'drowing' during Holocene submergence of barriers, shelf off Fire Island, New York. Bulletin of the Geological Society of America hbf86, 65–76.

SCHOFIELD, J.C. (1975): Sea level fluctuations cause periodic postglacial progradation, South Kaipara Barrier, North Island, New Zealand. New Zealand Journal of Geology and Geophysic **18**, 295–316.

Addresses of authors:
Christian Christiansen
Kristian Dalsgaard
Jens Tyge Møller
University of Aarhus
Department of Earth Sciences
Ny Munkegade Build. 520
8000 Aarhus C
Denmark
Dan Bowman
Ben-Gurion University of the Negev
Department of Geography
Beer Sheva 84105
P.O.B. 653
Israel

COASTAL DUNE CHRONOLOGY IN THE NORTH OF IRELAND

P. Wilson, Coleraine

Summary

Coastal dunes are abundant in the north of Ireland but a literature survey has revealed very few sites at which dune age is known. The earliest dunes pre-date 5,000 years B.P. and dune formation has occurred on several occasions since that time. More periods of dune construction are known for the last two millenia than the preceding years, and six dune building episodes have been recognised at Magilligan Foreland during this time interval. The causes of dune development are not fully understood, although within the last few centuries man's activities have undoubtedly played a prominent role. No clear regional trends in dune age have been identified.

1 Introduction

According to KINAHAN & Mc HENRY (1882) the area occupied by sand dunes and aeolian drift around the coast of Ireland is ca. 15,500 ha. Of this, ca. 12,000 ha are located in the six counties indicated on fig. 1, with ca. 6,700 ha in Donegal alone. However, despite this abundance of coastal aeolian sand, relatively few research papers have been devoted to explaining the geomorphological and chronological evolution of these deposits.

Coastal aeolian sand occurs predominantly as dunes but also as extensive near-level low-lying areas called machair (BASSETT & CURTIS 1985). The dunes generally comprise continuous or semi-continuous sand ridges orientated parallel to the coast. Young dune systems usually possess a distinctive fresh and sharp morphology while older dunes display more subdued relief. Sand thicknesses of up to 20 m are known to occur although most accumulations are considerably less than this. The deposits are composed of mixed quartz and carbonate sand that has a mean grain size range from ca. +1.9 to +2.7ø; within the medium and fine sand categories. Sorting coefficients range from 0.25 to 0.50ø, indicating well sorted and very well sorted sands. The carbonate content (mostly comminuted invertebrate shells) is usually <20%, although SHAW (1984) and BASSETT & CURTIS (1985) recorded some values in excess of 30%. Most Irish dunes are currently well-vegetated and stable.

At only a few sites in the north of Ireland is there detailed evidence available that enables us to establish the chronology of coastal dune development. Most of this evidence has been obtained during the past

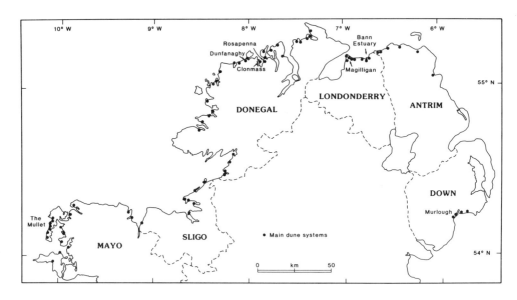

Fig. 1: *Coastal aeolian sand deposits in the north of Ireland. Sites mentioned in the text are indicated.*

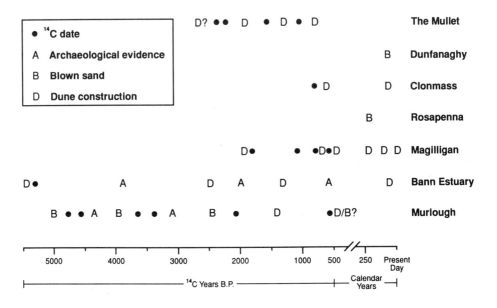

Fig. 2: *Phases of aeolian sand accumulation for seven sites in the north of Ireland. Archaeological evidence and ^{14}C dates, indicating surface stability, are also shown.*

15 years by investigations of buried soils and peats (CRUICKSHANK 1980, HAMILTON & CARTER 1983, SHAW 1984, WILSON & BATEMAN 1986, 1987, WILSON & FARRINGTON 1989) and through studies of recent coastal changes as documented by maps, aerial photographs and ground surveys (WILCOCK 1976, CARTER 1979, SHAW 1984, CARTER & WILSON in press). It is sad to report that although archaeological evidence from many sand dune systems indicates a long period of human occupation, extending as far back as the Mesolithic period in some instances, much of the early published work paid little heed to site stratigraphy with the consequent loss of potentially valuable information. Whilst accepting that stratigraphy may be a false guide to chronology in actively eroding dunes that contain evidence of human occupation (EVANS 1941, MAY & BATTY 1947, GIBBONS et al. 1988), a comprehensive multi-disciplinary study of these sites is required in order to increase our knowledge of coastal dune chronology.

The locations at which sand dune development can be assigned to particular periods of time are indicated on fig. 1. At some of these sites dune formation has occurred on several occasions since the mid-Holocene; at others the evidence is limited to dune building only in the last few centuries. These periods are depicted on fig. 2 along with archaeological evidence and ^{14}C dates that define phases of dune stability. The absence of early Holocene dunes is probably because sea-level was rising rapidly during this period (CARTER 1982) and any dunes that formed are likely to have been reworked.

2 Dune formation pre-2,000 years B.P.

Unequivocal evidence for dune formation prior to 2,000 years B.P. occurs in the dune systems of the Bann Estuary (fig. 1). From south of the estuary, HAMILTON & CARTER (1983) have described a silt-rich organic deposit interbedded in dune sands. Macro-fossils from the deposit yielded a ^{14}C date of $5,315 \pm 135$ years B.P. (UB-937), which is consistent with the pollen spectrum obtained. The date represents the earliest minimum age for Holocene dune development in Ireland (fig. 2). Buried and modern soil horizons in the overlying sands testify to three subsequent phases of dune building and stability, but these have not been dated.

The dunes on both sides of the Bann Estuary have long been known as sites of early human occupation and have produced a wealth of artifacts spanning the Neolithic to Medieval periods (GRAY 1879, KNOWLES 1887, HASSE 1890, COFFEY & PRAEGER 1904, BRUNICARDI 1914, HEWSON 1934, MAY & BATTY 1947). However, only COFFEY & PRAEGER (1904) and MAY & BATTY (1947) have provided stratigraphic details of the sites they investigated and from these it is evident that to the north of the estuary three phases of dune formation are represented; one pre-dating the Neolithic period and two post-dating the Bronze Age. The pre-Neolithic dunes are tentatively correlated with those recognised by HAMILTON & CARTER (1983) as pre-dating ca. 5,300 years B.P.

Older dunes may exist at Portrush, 8 km north-east of the Bann Estuary, where JESSEN (1949) observed low ridges of sand beneath peat of Boreal/Atlantic age and suggested that they formed as dunes in the early Holocene or an earlier period. The evidence for these sands being aeolian dunes is not wholly convincing. Unfortunately, coastal defence structures have obscured this section thus preventing further investigations that might resolve the issue.

At Murlough Nature Reserve in Co. Down (fig. 1), CRUICKSHANK (1980) has described buried podzolised soils that developed across a low-angle seaward-dipping sand sheet ranging in altitude from ca. 7 to 20 m O.D. The sand is ca. 6–7 m in thickness and is underlain by marine gravels. Radiocarbon dates and archaeological evidence indicate that sand deposition occurred prior to ca. 5,000 years B.P. (fig. 2). An aeolian origin for this sand is most likely but dunes (**sensu stricto**) are absent. A long period of land surface stability at Murlough from ca. 5,000 to 2,000 years B.P. is implied by several ^{14}C dates, archaeological finds and superimposed buried soil profiles. Double and triple profiles are contained within a 2 m thickness of sand which indicates that only minor amounts of sand were accumulating during this period. This land surface was buried by dunes sometimes between ca. 2,000 and 1,000 years B.P. The site provides a contrast to the Bann Estuary where dune formation is known to have occurred at least twice before 2,000 years B.P.

A period of dune construction prior to 2,000 years B.P. is also claimed by CROFTS (unpublished) for sites on and near the Mullet (figs. 1 & 2). Radiocarbon dates of 2,230±60 years B.P. (SRR-297) and 2,370±50 years B.P. (SRR-299) from charcoal within buried soil horizons were regarded as minimum estimates for the age of the underlying sands. The small thicknesses of sand beneath these dated horizons led CROFTS to suggest that dune development was followed by almost total erosion of the sands before pedogenesis commenced. Such events are difficult to substantiate and the potential problems associated with the interpretation of ^{14}C dates from charcoal have been outlined by WILSON & FARRINGTON (1989). There is clearly a need for more detailed study of these lower sands in order to determine their true nature and age.

3 Dune formation during the last two millenia

Many more phases and sites of dune formation are known for the last two millenia than the preceding years (fig. 2). CROFTS (unpublished), using ^{14}C dates and stratigraphic evidence, inferred dune building took place rapidly between four periods of stability and pedogenesis at the Mullet. At Clonmass (fig. 1), SHAW (1984) obtained a ^{14}C date of 800±40 years B.P. (SRR-2399) from a sub-dune organic-rich layer and also traced the development of a cuspate dune foreland since the beginning of the present century. On the basis of archaeological finds, a post-Iron Age/pre-Medieval period of dune formation is indicated north of the Bann Estuary (MAY & BATTY 1947) and WILCOCK (1976) has identified dunes formed within the last hundred years at the Bann mouth. Two phases of aeolian sand accumulation have also occurred at Murlough, the most recent of which post-dates ca. 600 years B.P. (CRUICKSHANK 1980).

The best documented location for dune formation in Ireland is Magilligan Foreland (figs. 1, 2 & 3), where three episodes of dune construction

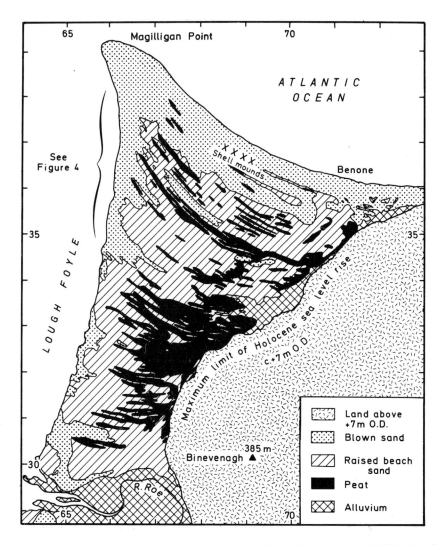

Fig. 3: *Holocene sediments of Magilligan Foreland (based on 1:50,000 Northern Ireland Geological Survey map) and location of landforms and sediments depicted in fig 4. Orientation and scale given by 1 km grid.*

have been recognised along the Lough Foyle shore (WILSON & BATEMAN 1986, 1987) and another three along the Atlantic shore (WILCOCK 1976, CARTER 1979, WILSON & FARRINGTON 1989, CARTER & WILSON in press). Magilligan is a triangular beach ridge plain of ca. 32 km² overlain by localised aeolian and lacustrine sediments. The sequence of beach ridges accreted diachronously northwards between the local maximum of the early Holocene eustatic transgression (ca. 7,000–6,500 years B.P.) and the re-

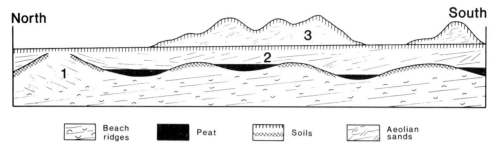

Fig. 4: *Landforms and sediments along a 3 km length of the Lough Foyle shore of Magilligan Foreland. Three phases of aeolian sand accumulation are indicated. Not to scale.*

turn to near present mean sea-level (ca. 2,000–1,500 years B.P.). As accretion of the beach ridge plain neared completion, prominent aeolian dunes formed on the seaward side of the beach ridges near to the present-day Magilligan Point. These dunes can be seen in cliff sections along the Lough Foyle shore (figs. 3 and 4), beneath more recent aeolian sands. The minimum age of the dunes is defined by a ^{14}C date of 1,790±50 years B.P (SRR-2438) from the base of a dune slack peat deposit that is laterally continuous with the dune soil. A stable land surface persisted until ca. 600 years B.P afterwhich the dunes were buried by aeolian sands. Radiocarbon dates obtained by WILSON & BATEMAN (1986) indicate that these sands migrated northwards from ca. 1,200 years B.P. in the south to ca. 600 years B.P. in the north. The planar surface displayed by these sands (fig. 4) may have resulted from erosion of a dune system or may be analogous to the machair sands described by BASSETT & CURTIS (1985). This planar surface is overlain by modern discontinuous dunes that WILSON & FARRINGTON (1989) suggest may have developed during the 17th and 18th centuries.

The dune system along the Atlantic shore of Magilligan consists of a series of shore-parallel ridges exceeding 1 km in width (fig. 3). Valves of *Arctica islandica*, taken from sub-dune storm-generated shell accumulations, yielded a ^{14}C age of 1,190±50 years B.P. (SRR-2439) corrected to ca. 780±60 years B.P. because of the natural deficiency of ^{14}C in British coastal waters (D.D. HARKNESS pers. comm. 1984). The date provides a tentative maximum age for the development of this dune system. More recent phases of dune formation along this shore have been recorded at Benone and Magilligan Point (fig. 3). WILCOCK (1976) has reported dune ridge development at Benone since 1851, while CARTER (1979) and CARTER & WILSON (in press) have described the evolution of a series of foredunes at Magilligan Point since the early 1950s. Up to ten dune ridges are present and have been dated to within one or two years by examination of air photographs and by ground surveys.

4 Causes of dune formation

Although a variety of factors are known to be of importance in the formation of coastal dunes (cf. RITCHIE 1972, PYE

1983), the causes of several phases of dune development in the north of Ireland are not yet firmly established. This is especially true for dunes formed prior to ca. 500 years B.P. It is not known, for example, whether these dunes represent:

a) primary phases of aeolian deposition associated with sea-level changes and sand supplied from beaches, or

b) the reworking of earlier, stabilized dune systems.

There is some evidence supporting the former hypothesis for the oldest dunes at the Bann Estuary, Murlough and Magilligan but it cannot be taken as conclusive. At these sites the earliest dunes are underlain by marine sands and gravels which indicate that sea-level probably fell before the aeolian sand accumulated. Dune construction related to falling sea-levels and the exposure of wide sand beaches to aeolian action has been reported from elsewhere in Europe (eg. CHRISTIANSEN & BOWMAN 1986).

Several of the dune systems formed within the last 500 years are either known or thought to have developed in response to man's activities in the coastal zone. These activities have resulted primarily in the remobilisation of previously stabilized dunes. From documentary evidence, QUINN (1977) has summarised the most recent episodes and causes of sand blowing in Donegal, Sligo and Mayo. In the late 17th and early 18th centuries, the village of Rosapenna (fig. 1) was overwhelmed by sand blown from dunes into which rabbits had been introduced and allowed to multiply. On dunes near Dunfanaghy (fig. 1) the sand-binding marram grass was cut extensively during the First World War, resulting in large-scale sand movements and the eventual abandonment of local houses. At sites in Sligo and Mayo, burning and overgrazing of the marram during the 19th and early 20th centuries led to villages and agricultural land being engulfed by sand.

Rabbits have also been considered responsible for sand erosion and dune development at Magilligan. WILSON & FARRINGTON (1989) suggested that the modern discontinuous dunes along the Lough Foyle shore (unit 3, fig. 4) may have originated during the 17th and 18th centuries when Magilligan contained Ireland's largest rabbit warren.

Further evidence linking dune building to man's activities is provided by WILCOCK (1976) and SHAW (1984). At the mouth of the Bann, dunes have extended considerably in the last hundred years following the construction of estuary-mouth training walls. These new dunes are strongly ridged in contrast to the older dunes and sand accumulation was noted as continuing into the 1970s (WILCOCK 1976). Land reclamation in the upper Clonmass estuary was regarded by SHAW (1984) as a possible cause of dune erosion/construction in the lower part of the estuary. A reduction of the tidal prism as a result of land-taking was thought to have caused migration of the ebb channel, allowing erosion of pre-existing dunes and the growth of a new dune system.

The factors responsible for the development of other recent dune systems are not as easy to identify. Foredune growth at Magilligan Point since the early 1950s (CARTER 1979, CARTER & WILSON in press) seems to be part of a 35-40 year cycle of coastal expansion and contraction that has been evident since the early 19th century. The cause of this cycle is still conjectural but may be related to sequential climatic change, variations in near-shore wave refraction or, perhaps, to land reclamation in the upper Foyle Estuary.

5 Conclusions

The abundance of coastal aeolian sand in the north of Ireland has stimulated relatively little detailed research into the origins, characteristics and age of the deposits. The chronological evidence presented above is drawn from only a few sites/studies and further investigations are required in order to establish whether the phases of dune building identified are regionally, as opposed to locally, significant. However, the search for regional trends in dune chronology may not be appropriate on a coast displaying such strong morphological contrasts and also where marked variations in wave energy occur, between east, north and west. This brief survey of the available literature has highlighted numerous localised phases of dune construction but no clear regional trends have emerged.

Acknowledgements

The author thanks R.W.G. Carter for constructive criticism of the manuscript, R.S. Crofts for access to unpublished data and K. McDaid for preparing the diagrams.

References

BASSETT, J.A. & CURTIS, T.G.F. (1985): The nature and occurrence of sand-dune machair in Ireland. Proceedings of the Royal Irish Academy **85B**, 1–20.

BRUNICARDI, Mrs. (1914): The shore-dwellers of ancient Ireland. Journal of the Royal Society of Antiquaries of Ireland **4** (6th series), 185–213.

CARTER, R.W.G. (1979): Recent progradation of the Magilligan Foreland, Co. Londonderry, Northern Ireland. Publications du CNEXO: Actes de Colloques **9**, 17–27.

CARTER, R.W.G. (1982): Sea-level changes in Northern Ireland. Proceedings of the Geologists' Association **93**, 7–23.

CARTER, R.W.G. & WILSON, P. (in press): Geomorphological, sedimentological and pedological influences in coastal foredune development at Magilligan, Northern Ireland. In: K.F. Nordstorm, N.P. Psuty & R.W.G. Carter (eds.), Coastal dunes: processes and morphology. Wiley, Chichester.

CHRISTIANSEN, C. & BOWMAN, D. (1986): Sea-level changes, coastal dune building and sand drift, north-western Jutland, Denmark. Geografisk Tidsskrift **86**, 28–31.

COFFEY, G. & PRAEGER, R.L. (1904): The Antrim raised beach: a contribution to the Neolithic history of the north of Ireland. Proceedings of the Royal Irish Academy **25C**, 143–200.

CRUICKSHANK, J.G. (1980): Buried, relict soils at Murlough sand dunes, Dundrum, Co. Down. Irish Naturalists' Journal **20**, 21–31.

EVANS, E.E. (1941): A sandhill site in Co. Donegal. Ulster Journal of Archaeology **4** (3rd series), 71–75.

GIBBONS, M.A., HIGGINS, J. & McCORMICK, F. (1988): Truska midden sites. In: M. O'Connell & W.P. Warren (eds.), Connemara. 65–68. Field Guide **11**, Irish Association for Quaternary Studies.

GRAY, W. (1879): The character and distribution of the rudely-worked flints of the north of Ireland, chiefly in Antrim and Down. Journal of the Royal Historical and Archaeological Association of Ireland **5** (4th series), 109–143.

HAMILTON, A.C. & CARTER, R.W.G. (1983): A mid-Holocene moss bed from eolian dune sands near Articlave, Co. Londonderry. Irish Naturalists' Journal **21**, 73–75.

HASSE, L. (1890): Objects from the sandhills at Portstewart and Grangemore, and their antiquity. Proceedings and Papers of the Royal Society of Antiquaries of Ireland **1** (5th series), 130–138.

HEWSON, L.M. (1934): Notes on Irish sandhills. Journal of the Royal Society of Antiquaries of Ireland **4** (7th series), 231–244.

JESSEN, K. (1949): Studies in late Quaternary deposits and flora-history of Ireland. Proceedings of the Royal Irish Academy **52B**, 85–290.

KINAHAN, G.H. & McHENRY, A. (1882): A handy book on the reclamation of wastelands, Ireland. Hodges, Figgis & Co., Dublin.

KNOWLES, W.J. (1887): The prehistoric sites of Portstewart, County Londonderry. Journal of the Royal Historical and Archaeological Association of Ireland **8** (4th series), 221–237.

MAY, A. McL. & BATTY, J. (1947): The sandhill cultures of the River Bann estuary, Co. Londonderry. Journal of the Royal Society of Antiquaries of Ireland **78**, 130–156.

PYE, K. (1983): Coastal dunes. Progress in Physical Geogaphy **7**, 531–557.

QUINN, A.C.M. (1977): Sand dunes: formation, erosion and management. An Foras Forbartha, Dublin.

RITCHIE, W. (1972): The evolution of coastal sand dunes. Scottish Geographical Magazine **88**, 19–35.

SHAW, J. (1984): Clonmass Estuary. In: P. Wilson & R.W.G. Carter (eds.), Northeast Co. Donegal and Northwest Co. Londonderry. 10–21. Field Guide **7**, Irish Association for Quaternary Studies.

WILCOCK, F.A. (1976): Dune physiography and the impact of recreation on the north coast of Ireland. Unpublished D. Phil. thesis, The New University of Ulster.

WILSON, P. & BATEMAN, R.M. (1986): Nature and paleoenvironmental significance of a buried soil sequence from Magilligan Foreland, Northern Ireland. Boreas **15**, 137–153.

WILSON, P. & BATEMAN, R.M. (1987): Pedogenic and geomorphic evolution of a buried dune palaeocatena at Magilligan Foreland, Northern Ireland. CATENA **14**, 501–517.

WILSON, P. & FARRINGTON, O. (1989): Radiocarbon dating of the Holocene evolution of Magilligan Foreland, Co. Londonderry. Proceedings of the Royal Irish Academy **89B**, 1–23.

Address of author:
Peter Wilson
Department of Environmental Studies
University of Ulster at Coleraine
Cromore Road
Coleraine
Co. Londonderry BT52 1SA
Northern Ireland
U.K.

NEW PUBLICATION

Arnt Bronger & John A. Catt (Editors)

PALEOPEDOLOGY
NATURE AND APPLICATION OF PALEOSOLS

CATENA SUPPLEMENT 16

hardcover/240 pages/numerous figures, photos and tables

ISSN 0936-2568/ISBN 3-923381-19-0

list price: DM 139.-/US $ 75.-/standing order price CATENA SUPPLEMENTS: DM 97,30/US$ 52.50

ORDER FORM

☐ Please send me .... at the rate of DM 139.-/ US $ 75.- .... copies of CATENA SUPPLEMENT 16.

☐ I want to subscribe to CATENA SUPPLEMENTS (30% reduction on the list price) starting with no..

Name ....

Address ....

Date ....

Signature: ....

Please charge my credit card: ☐ MasterCard/Eurocard/Access ☐ Visa ☐ Diners ☐ American Express

Card No.: .... Expiration date: ....

Please, send your orders to:

CATENA VERLAG, Brockenblick 8, D-3302 Cremlingen-Destedt, West Germany, tel.05306-1530, fax 05306-1560

USA/Canada:**CATENA VERLAG**, Attn. Denize Johnson, P.O.Box 368, Lawrence, KS 66044, USA, Tel. (913) 843-1234, fax (913) 843-1244

THE CHRONOLOGY OF COASTAL DUNE DEVELOPMENT IN THE UNITED KINGDOM

M.J. Tooley, Durham

Summary

The development of inland and coastal dunes in the United Kingdom is described, after a brief review of the methodologies that have been used to calculate the age of the dunes. Attention is drawn to the insufficiency of the data, and, in particular, their unevenness spatially and temporally: more data have been collected from the dunes of north-west England, the machair of north-west Scotland and the dunes of northern Ireland than from other parts of Britain. The data from the United Kingdom are compared with the Older Dune and Younger Dune sequences of The Netherlands.

1 Introduction

The coastal dunes of Great Britain occupy an area of 56,300 ha (DOODY 1989, 53). This does not include the dunes of Northern Ireland, some of which, such as those of Magilligan, Co. Londonderry, occupy an area of 2000 ha (CARTER 1982, 15). In addition, the figure underestimates the total area, part of which has been built up and part is the landward margin of the dunes, where in some cases the sand has attenuated and been incorporated into a peaty soil.

Some attempts have been made to date the dunes based on rates of accumulation, the depletion of the calcium carbonate content of dune soils, the cartographic and documentary record and the archaeological record.

READE (1881, 438) measured the rate of sand accumulation at two sites on the coast of south-west Lancashire over a number of years and calculated the annual accumulation rate. He measured the extent of the dune field in south-west Lancashire and from an assumed sand thickness calculated the age of the dunes to 2580 years. SALISBURY (1920, 322) used Speed's map of 1610, an estate map of 1736 and the annual rings of *Salix repens* in the oldest dune slacks to calculate the average rate of coastal progradation and dune advance along a transect from the embryo dunes to the heather covered dunes near Ainsdale in south-west Lancashire (fig. 1). This methodology was extended to other dune systems, such as Braunton Burrows (SALISBURY 1952, 299).

Archaeological and documentary sources have been employed by many investigators to establish the chronology of dune systems (for example HIGGINS

1933, EDLIN 1976, CRUICKSHANK 1980, CHERRY 1982, CHERRY & CHERRY 1986, BELL 1987), whilst STEERS (1959) has reviewed the archaeological and historic record of dune movement in Britain.

Unfortunately, no conspectus of dune chronology exists for the United Kingdom, and the literature is both extensive and scattered. No coherent methodology, such as that employed in the Netherlands (JELGERSMA et al. 1970), has been applied, and conclusions must be site based.

In this contribution, some of the evidence for dunes from the late-glacial and post-glacial in the United Kingdom is presented. For the latter period, details of dune development from north-west England are given.

2 Dune development during the Late Devensian Age

There is some evidence that there was a marine accompaniment to deglaciation in the Irish Sea basin, and Late Devensian age brackish water and marine sediments have been recorded southeast of the Isle of Man and on land near the coast of south-west Lancashire (TOOLEY 1985a, 98). Along the Late Devensian shore, sand dunes may have occurred for the pollen record from Moss Lake, Liverpool (GODWIN 1959, 136) contains a grain of *Agropyron junceiforme*, the sand couch grass, which is referred to the Allerød interstadial. *A. junceiforme* is common today along the coast of south-west Lancashire from Crosby to Hesketh (SAVIDGE et al. 1963, 324), but it extends north to about 63°N where it is found on embryo dunes.

During the succeeding Younger Dryas period, cover sands accumulated in south-west Lancashire. The sand occupied 200 km^2 and extends from the present coast some 20 km inland, and is recorded at altidues ranging from -14 m Ordnance Datum to $+120$ m OD in the east. The Shirdley Hill Sand Formation, named after Shirdley Hill, northwest of Clieves Hills (fig. 1), is made up of fine to medium sands, with augite dominating the heavy mineral assemblage and indicating a fluvioglacial sediment source for the sands (WILSON et al. 1981, 222). East of Downholland Moss, the sand takes the form of low parabolic dunes, overlapped on their flanks by peat. The sand rarely exceeds 5 m in thickness, and is usually a venner of 1 m or less: greater thicknesses occur where there are depressions in the underlying till. Such an example can be found at Clieves Hills (fig. 2, section 1), where peat growth ended shortly after 10,455±110 (Hv. 4710) when sand accumulated at the site (TOOLEY 1978, 142, INNES et al. 1989, 65). Elsewhere, Shirdley Hill Sand has been recorded beneath the peat, and together they have been strongly cryoturbated (TOOLEY 1985a, 111). The Shirdley Hill Sand is a true coversand that was reworked by multidirectional winds, which explains the uniformity of the particle size distribution over such an extensive area (WILSON et al. 1981, 228). However, it was further reworked by water and wind, especially along the landward margins of the formerly extensive lakes such as Martin Mere (TOOLEY 1985, 111) in the tidal flat and lagonal zone of south-west Lancashire. There is also some evidence of reactivation during the late mesolithic period following disturbance of the vegetation cover by fire: at Firswood Road, peat began to accumulate on the sand as

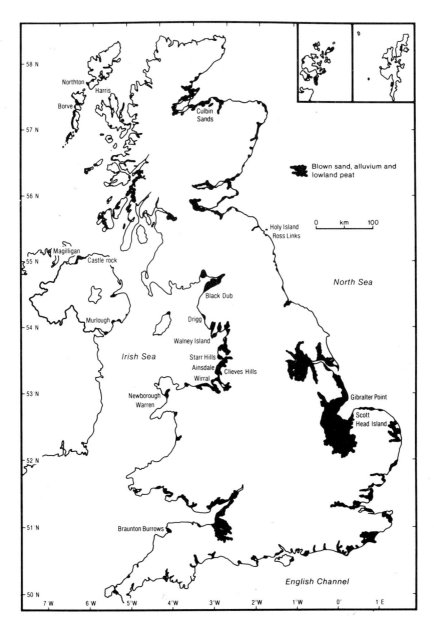

Fig. 1: *A map of the British Isles showing the distribution of blown sand, marine alluvium and lowland peat. The sand dunes mentioned in the text are named on the map.*
Based on: International Quaternary Map of Europe, Sheet 6 København, 1:2,500,000 Hannover 1970; Quaternary Map of the United Kingdom, North and South, 1:625,000, Institute of Geological Sciences; The Atlas of Britain and Northern Ireland, p. 18. Superficial deposits 1:2,000,000.

late as 6195±80 (Hv. 4711).

3 Dune Development during the Flandrian Age

The Shirdley Hill Sand Formation was only marginally and indirectly affected by sea level changes, whereas the dunes of Flandrian Age around the coasts of the United Kingdom are invariably associated with tidal flat and lagoonal sediments and have been affected by positive and negative sea-level tendencies and by climatic change.

The oldest dune slack peat recorded in the United Kingdom comes from the Black Dub on the Cumbrian coast. Here interdigitating bands of clay, peat and sands are exposed along the south side of the stream channel, which has cut through the sediments of the "25 ft" raised beach. Immediately south is a low dune. Pollen analysis of the peat showed an assemblage rich in halophytes, such as Chenopodiaceae, *Artemisia* and *Plantago cf. maritima*. The tree pollen assemblage indicated a Flandrian Ic chronozone age and this was confirmed by a radiocarbon assay of 8480±205 (Hv. 5224) (HUDDART et al. 1977, 139).

This date appears to be quite exceptional, and most dates from organic material subjacent to dune sand indicate an inception of dune formation towards the end of Flandrian Chronozone II and during Flandrian Chronozone III.

In Scotland, at Borve on Benbecula, Outer Hebrides, RITCHIE (1966, 85) has demonstrated that sand deposition began about 5700±170 (I.1543) and continued intermittently at Northton, Isle of Harris until 3481±54 BP (BM 707) (RITCHIE 1979, 116). In Northern Ireland, the earliest date for the formation of sand dunes is 5315±135 at Castlerock (CARTER 1982, 15). Whereas in England, apart from the evidence from the Black Dub, the indications from the Lancashire dunes are that dune formation is younger than the elm decline, that is younger than 5000 BP.

At Downholland Moss (fig. 2, sections 2 and 3) east of Ainsdale, blown sand up to a metre thick has overblown the peat and forms the landward limit of the dunes. A radiocarbon assay on peat immediately subjacent to the sand yielded a date of 4090±170 (Hv. 4705). Blown sand is incorporated into the surface peat further east: for example at section 3 on fig. 2, 25% of the upper stratum is sand, which levens the limnic peat, and indicates that sand was blowing into a lagoon at this time, whilst about 1 km seaward a thick stratum of blown sand was being laid down. Latterly, the sand was overblowing characteristic tidal flat and lagoonal sediments that accumulated on Downholland Moss from about 8000 BP and terminated towards the end of Flandrian Chronozone II.

Further north, in south-west Cumbria, occupation layers of neolithic age have been recorded above shingle ridges and beneath dune sand at the North End of Walney Island (TOOLEY 1978, 144). In Northern Ireland, a neolithic hearth has been described from the Murlough Sand Dunes on Dundrum Bay, and charcoal from the hearth has been dated to 4775±140 (UB 412) and 4565±135 (UB 413) (CRUICKSHANK 1980, 27).

All these data point to a period of dune building in north-west England, north-west Scotland and Northern Ireland towards the end of Flandrian Chronozone II and the beginning of Flandrian Chronozone I.

Although there are some indications

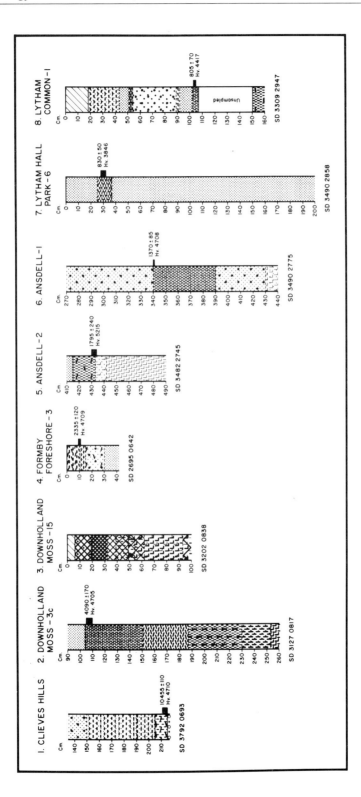

Fig. 2: The stratigraphy from inland and coastal dunes in Lancashire, north-west England. The stratigraphic signatures are according to the scheme proposed by TROELS-SMITH (1955).

of cycles of stability and instability in the succeeding millennium, notably from dunes to Co. Down and Co. Donegal in northern Ireland (CARTER 1982, 19) and at Drigg where blown sand overlies peat containing a hearth, charcoal from which has been dated to 3780±55 (UB 905) and 4135±55 (UB 906) (CHERRY 1982, 5), there appears to be a marked and widespread stage of dune stability between 2500 and 2000 BP when soils developed, dune slack peat accumulated and occupation of the dunes occurred. In Lancashire, this stage is manifest on Formby Foreshore by a dune slack soil with woody detritus dated to 2335±120 (Hv. 4709) (TOOLEY 1978, 144) located in the upper intertidal zone and shown as section 4 in fig. 2. At a lower altitude in the intertidal zone at Formby, the soil has been poached by cattle, the hoof prints of which are clearly visible and are those of domestic ox (Dr. J. JEWEL, personal communication, 16 September 1974), on the Wirral, to the south-west, KENNA (1986, 14) has dated an impersistent thin black clay peat underlaid by sand of marine or estuarine origin and overlaid by peaty sand that appears to have formed under dune slack conditions. In Northern Ireland, there are five radiocarbon dates that fall within this period of stability. In Northumberland, this period of stability ended shortly after 2810±40 when sand began to accumulate at Low Hauxley (INNES & FRANK 1928).

In Lancashire, a further period of stability is marked by organic horizons and peat intercalating sand dated to between 1795±240 (Hv. 5214) and 1370±85 (Hv. 4708): the stratigraphy is shown as sections 5 and 6 in fig. 2. At the junction of the peat with the overlying sand, near section 5, a left mandibular ramus, that had been gnawed, was recovered, and identified as the jawbone of a young ox (Dr. J. JEWELL, personal communiaction, 24 August 1971). These dates may indicate merely an extension of the preceding period of stability, but taken with the data from northern Ireland (CARTER 1982, 19), the clustering of dates from organic soil, peat and shells points to a discrete period of stability.

The final period of stability, both preceded by and succeeded by sand blowing, occurred about 800 BP, and sections 7 and 8 on fig. 2 from the Lytham sand dunes (the Starr Hills) provide the evidence for this. The stability of the Lytham sand dunes coincides with the foundation of the Benedictine monastery of Lytham — a cell of the monastery at Durham — the records of which from the fifteenth and sixteenth centuries provide evidence of priods of dune stability when an income was derived from the dune pastures and of dune instability when the pastures were overblown by sand and there was no income (TOOLEY 1985b, 225). On the Wirral (KENNA 1986, 15), on the Norfolk coast (FUNNELL & PEARSON 1984, 125), on the Northumberland coast (FRANK 1982, 24) and in northern Ireland there are comparable dates such as the date 800±40 (SRR. 2399) from the dune at the mouth of the Clonmass estuary (CARTER et al. 1989, 21), bearing witness to a widespread period of stability during the period of the Medieval Warm Epoch.

4 Discussion

In the United Kingdom there are about 50 radiocarbon dates that can be used to establish a chronology of the coastal and inland sand dunes. These dates are scattered both geographically and temporally, and, although there is some clustering of the data are insufficient to come to firm conclusions.

For Western Europe, the most detailed dune chronology has been worked out by JELGERSMA et al. (1970), and some comparison is appropriate. The oldest recorded Older Dunes in the Netherlands formed after 4800 BP on the most landward coastal barrier and was completed by 4100 BP (op.cit., 98). This appears to correlate with the oldest and most landward of the dunes on the Lancashire coast, but is much younger than the inception of dune building in north-west Scotland and northern Ireland.

The phases of dune stability from 2500–2000 BP in the United Kingdom occur during Older Dune II in the Netherlands, but there are insufficient data to permit the resolution into sub-stages B1 to B3.

The final period of stability about 800 BP is widespread, and correlates with the non-active dune phase, during Phase I and Phase II of the formation of the Younger Dunes in the Netherlands (KLIJN 1990, this volume).

5 Conclusions

The chronology of inland and coastal sand dunes in the United Kingdom is fragmentary and derives from dates obtained from archaeological or sea-level investigations. Nevertheless, some temporal patterns do emerge and these can be correlated with the more coherent chronology from the Netherlands. It is clearly important that site-based projects devoted to the establishment of a sand dune chronology in the British Isles are undertaken to resolve general problems of sea-level and groundwater level changes, climatic changes and specific problems of vegetation succession, soil development and prehistoric and historic occupation of dune areas. Furthermore, the volume of sand reservoirs and increments to these reservoirs need to be calculated (as members of the Dutch Geological Survey are undertaking (S. JELGERSMA, personal communication)) in order to evaluate the impacts of changing sea-level on the sand budget.

References

BELL, M. (1987): Recent molluscan studies in the south west. In: N.D. Balaam, B. Levitan & V. Straker (eds), Studies in palaeoeconomy and environment in south-west England. British Archaeological Reports, Oxford, British Series **181**, 1–8.

CARTER, R.W.G. (1982): Sea-level change in Northern Ireland. Proceedings of the Geologist's Association **93** (1), 7–23.

CARTER, R.W.G., DEVOY, R.J.N. & SHAW, J. (1989): Late Holocene Sea levels in Ireland. Journal of Quaternary Science **4** (1), 7–24.

CHERRY, J. (1982): Sea cliff erosion at Drigg, Cumbria: evidence of prehistoric habitation. Transactions of the Cumberland and Westmorland Antiquarian and Archaeological Society **83**, 1–6.

CHERRY, J. & CHERRY, P.J. (1986): Prehistoric habitation sites in West Cumbria: Part IV. The Eskmeals area. Transactions of the Cumberland and Westmorland Antiquarian and Archaeological Society **86**, 1–17.

CRUICKSHANK, J.G. (1980): Buried, relict soils at Murlough sand dunes, Dundrum, Co. Down. Irish Naturalists' Journal **20** (1), 21–31.

DOODY, P. (1989): Conservation and development of the coastal dunes in Great Britain. In: F. van der Meulen, P.D. Jungerius & J.H. Visser (eds.), Perspectives in coastal dune management, 53–67. SPB Academic Publishing bv., The Hague.

EDLIN, H.L. (1976): The Culbin Sands. In: J. Lenihan & W.W. Fletcher (eds.), Reclamation, 1–13. Blackie, Glasgow.

FRANK, R. (1982): A Holocene peat and dune sand sequence on the coast of northeast England

— a preliminary report. Quaternary Newsletter **36**, 24–32.

FUNNELL, B.M. & PEARSON, I. (1984): A guide to the Holocene geology of North Norfolk. Bulletin of the Geological Society of Norfolk **34**, 123–140.

GODWIN, H. (1959): Studies of the post-glacial history of British vegetation XIV. Late-glacial deposits at Moss Lake, Liverpool. Philosophical Transactions of the Royal Society, Series B, **242 (698)**, 127–149.

HIGGINS, L.S. (1933): An investigation into the problem of the sand dune areas on the south Wales coast. Archaeologia Cambrensis **88**, 26–67.

HUDDART, D., TOOLEY, M.J. & CARTER, P.A. (1977): The coasts of northwest England. In: C Kidson & M.J. Tooley (Eds.), The Quaternary History of the Irish Sea. 119–154. Liverpool, Seel House Press.

INNES, J.B. & FRANK, R.M. (1988): Palynological evidence for Late Flandrian coastal changes at Druridge Bay, Northumberland. Scottish Geographical Magazine **104 (1)**, 14–23.

INNES, J.B., TOOLEY, M.J. & TOMLINSON, P.R. (1989): A comparison of the age and paleoecology of some sub-Shirdley Hill Sand peat deosits from Merseyside and south-west Lancashire. The Naturalist **114**, 65–69.

JELGERSMA, S., DE JONG, J., ZAGWIJN, W.H. & VAN REGTEREN, J.F. (1970): The coastal dunes of teh western Netherlands; geology, vegetation history and archaeology. Mededelingen Rijks Geologische Dienst **21**, 93–164.

KENNA, R.J.B. (1986): The Flandrian sequence of north Wirral (N.W. England). Geological Journal **21**, 1–27.

KLIJN, J.A. (1990): The younger dunes in The Netherlands; chronology and causation. This volume.

READE, T.M. (1881): The date of the last change of level in Lancashire. The Quarterly Journal of the Geological Society of London **37 (147)**, 436–439.

RITCHIE, W. (1966): The postglacial rise in sea-level and coastal changes in the Uists. Transactions of the Institute of British Geographers **39**, 79–86.

RITCHIE, W. (1979): Machair development and chronology in the Uists and adjacent islands. Proceedings of the Royal Society of Edinburgh 77B, 107–122.

SALISBURY, E.J. (1920): Note on the edaphic succession in some dune soils with special reference to the time factor. Journal of Ecology **8**, 322–328.

SALISBURY, E.J. (1952): Downs and Dunes: their plant life and its environment. London, G. Bell and Sons, Ltd.

SAVIDGE, J.P., HEYWOOD, V.H. & GORDON, V. (1963): Travis' Flora of South Lancashire. Liverpool, The Liverpool Botanic Society.

STEERS, J.A. (1959): Archaeology and physiography in coastal studies. In: Second Coastal Geography Conference, Louisiana, 317–340.

TOOLEY, M.J. (1978): Sea-level Changes: north west England during the Flandrian Stage. Oxford, Clarendon Press.

TOOLEY, M.J. (1985a): Sea-level changes and coastal morphology in North-west England. In: R.H. Johnson (ed.), The geomorphology of North-west England. Manchester University Press, 94–121.

TOOLEY, M.J. (1985b): Climate, sea-level and coastal changes. In: M.J. Tooley & G.M. Sheail (eds.), The Climatic Scene. London. George Allen and Unwin, 206–234.

WILSON, P., BATEMAN, R.M. & CATT, J.A. (1981): Petrography, origin and environment of deposition of the Shirdley Hill Sand of south-west Lancashire, England. Proceedings of the Geologists' Association **92 (4)**, 211–229.

Address of author:
M.J. Tooley
Department of Geography
University of Durham
Science Laboratories
South Road
Durham DH1 3LE
UK

THE YOUNGER DUNES IN THE NETHERLANDS; CHRONOLOGY AND CAUSATION

J.A. **Klijn**, Wageningen

Summary

The so-called Younger Dunes in The Netherlands originate largely from secondary dune formation. Three distinct phases can be distinguished based on C^{14}- and historical datings and additional geomorphological data on migration speeds of dunes. Phase I must have begun between 800 and AD 1000 and ended between AD 1200 and 1300. Phase II occurred approximately between AD 1300 and 1600 and Phase III between AD 1750 and 1850. The landward migration of dunes is considered to be primarily triggered by coastal erosion. Coastal erosion is assumed to have been induced by a rising sea level and an increased storm surge frequency, both mechanisms being governed by climatic changes. Data on climate history, storm surge frequency, coastal erosion and phases in dune formation seem to correlate.

1 Introduction

Extensive sections of the Dutch North Sea coast are fringed by coastal dunes. These dunes can be distinguished by the so-called Beach Barrier and Older

ISSN 0722-0723
ISBN 3-923381-23-9
©1990 by CATENA VERLAG,
D–3302 Cremlingen-Destedt, W. Germany
3-923381-23-9/90/5011851/US$ 2.00 + 0.25

Dune deposits and the Younger Beach and Dune deposits. The first group includes parallel beach barriers and related beach plains developed from 2800 until 1500 B.C. (VAN STRAATEN 1965), in the western part of the Netherlands even up to Roman times (ROEP 1984). Dune development took place concomittantly and continued into the early Middle Ages and even later on the barrier islands in the northern part of the country, until approx. AD 1200 (VAN STAALDUINEN 1977, DE JONG 1984).

Beach barriers, related older dunes and beach plains situated in the west, are partly covered by Younger Dunes. These dunes originate from about the High Middle Ages. Their sudden formation is considered to be one of the most catastrophic events in coastal history. The Younger Dunes show a distinct relief in contrast to the Older Dunes and consist largely of so-called secondary dunes such as parabolic dunes, comb dunes, precipitation ridges, secondary barchans and deflated (secondary) dune slacks (KLIJN 1981). Primary dunes, i.e. (former) foredunes and cut-off beach plains forming dune slacks are mainly restricted to the estuarine coast in the southwest and the (former) "wadden" coast in the north (KLIJN 1981). The latter can historically be dated as younger than AD 1570. Their origin is connected to local coastal progradation common in estuarine and barrier island environments. Secondary dune systems are much more widespread in The Netherlands (90% of the area of the Younger Dunes). Their dating is based on C^{14}, archeological and at times historical data. JELGERSMA et al. (1970) characterized three main phases in (Younger) dune formation in the mainland dunes: Phase I(a+b): ca. AD 1100–1300, Phase II: AD 1400–1600 and Phase III: ca. AD 1700–1850.

Phase I (a & b) is described as a phase (a) in which the former Older Dunes were levelled, followed by the deposition of thin horizontal beds (b). In Phase I(a) a precipitation ridge was also formed at the landward boundary. A large scale parabolization, occurred in Phase II. Phase III witnessed smaller scale parabolization near the coastline and in a few other areas susceptible to erosion. In the next section datings are compiled and reinterpreted from a geomorphological point of view.

2 Dating of the formation of the Younger Dunes

Datings of coastal dune sediments, relevant to the Younger Dunes, based on C^{14} analysis in the secondary mainland dunes of The Netherlands are mentioned by JELGERSMA et al. (1970), DE MULDER et al. (1983) and ZAGWIJN (1984); datings relating to the wadden island of Terschelling by VAN STAALDUINEN (1977) and DE JONG (1984). Material relevant to C^{14}-datings for the start of the Younger Dune-formation was taken from organic soil horizons or peat layers on top of the Older Dunes or from humic layers in the lower part of the Younger Dune deposits. Tab. 1 shows uncorrected as well as corrected C^{14} ages.

From geomorphological evidence (JELGERSMA et al. 1970, KLIJN 1981), it is clear that the Younger Dunes from Phase I developed into secondary dunes, which must have migrated from the coast landwards. Therefore, the distance (in transport direction) to the original (!) coastline as well as migration speed (tab. 2) are essential parameters in order to calculate the transport time. By means of data on migration speed, substantial corrections for the start of the dune forming phases were suggested by KLIJN (1981) and later by ZAGWIJN (1984).

2.1 Phase I

In Phase I, dune migration must have been rather rapid, after having taken into account sedimentological and geomorphological features. A yearly migration speed of 10 to 20 m was assumed by KLIJN (1981). This assumption, together with a transport distance varying from app. 2 to 4 km supported the conclusion that the very start of this phase near the coast could have been one to four centuries earlier than AD 1100 as mentioned by JELGERSMA et al. (1970). KLIJN assumed that the process could have commenced in the 9th century. From the oldest written sources RENTENAAR (1977) decided that enormous sand blowing halfway in the 10th century could be considered to be the first phase of Younger Dune formation. ZAGWIJN (1984), referring to rather high velocities for small migrating desert dunes (25 m/yr) calculated that the start of the process had to be 140 to 180 years before the datings established by JELGERSMA et al. (1970) at locations 3500 to 4500 meters respectively inland.

Nr. of sample	Location	Situation in profile	Conventional C-14 age	Corrected C-14 age * SUESS 1969, STUIVER 1982 **	Distance to present coast line
GrN* 4480	Velsen Hoogovens II	Peat layer 1 m above boundary O.D./Y.D. (4.5 m +NAP)	1010±45	ca. 940±45 (= 1010±45) A.D.	ca. 3 km
GrN* 4563	Velsen Hoogovens II	in Y.D.I: 2.25 m above boundary O.D./Y.D. (5.75 m +NAP)	940±45	ca. 900±45 (= 1050±45) A.D.	ca. 3 km
GrN* 4561	Velsen Hoogovens III	Boundary O.D./Y.D. (3.75 m +NAP)	1370±70	ca. 1330±70 (= 620±70) A.D.	ca. 3 km
GrN* 4118	Velsen Hoogovens III	Peat layer in Y.D.I; 1.5–2 m above boundary O.D./Y.D. (5.6 m +NAP)	810±70	ca. 730±70 (= 1220±70) A.D.	ca. 3 km
GrN* 5040	Velsen Vormenhal	Boundary O.D./Y.D. (3.75 m +NAP)	1090±35	ca. 920±35 (= 1030±35) A.D.	ca. 1 1/4 km
GrN* 4564	Amsterdamse Waterleidingduinen A-VII	In Y.D. Ib: ca. 0.5 m above boundary O.D./Y.D. (5.50 m +NAP)	850±55	ca. 730±35 (= 1220±35) A.D.	ca. 2 1/2 km
GrN* 4664	Amsterdamse Waterleidingduinen A-VII	Some dm above boundary O.D./Y.D. (4.90 m +NAP)	820±50	ca. 720±35 (= 1230±35) A.D.	ca. 2 1/2 km
GrN* 4642	Amsterdamse Waterleidingduinen	Trunk Quercus boundary O.D./Y.D.	860±40	ca. 730±40 (= 1220±40) A.D.	ca. 2 1/2 km
GrN* 5237	Egmond Watertoren	Boundary O.D./Y.D.??	1090±45	ca. 920±35 (= 1030±35) A.D.	ca. 1/2 km
GrN** 6447	Midden Heerenduin	Boundary Y.D. Top peat layer	970±45	ca. 1030±45 920±45 A.D.	ca. 3 km?
GrN** 12085	Kijkduin	Boundary Y.D. Top peat layer	1105±35	1105±35 900–1000 A.D.	ca. 0 km?

Tab. 1: *Uncorrected and corrected C-14 datings of Younger Dunes (YD) in the western part of the Netherlands. OD = Older Dunes; NAP = Ordnance Datum Netherlands.*

He considers Phase Ia to have been situated between AD 1000 and 1180. Phase Ib should in his opinion have taken place between AD 1180 and 1330.

2.2 Phase II

Phase II was a period of extensive parabolic dune forming. Assuming migration speeds of 5 to 10 m/yr usual for parabolic dune formation (tab. 2), KLIJN (1981) calculated a duration of at least three centuries. Since Phase II can safely be considered to have ended around 1600, in view of undisputed evidance from historical maps and other sources, it must have begun around AD 1300 or before (JELGERSMA et al. 1970, KLIJN 1981). ZAGWIJN (1984) assumed rather high velocities (10 m/yr) and deduced that this phase ought to

Barchane/Free migrating dunes	Authors/Country
25 m/y	VAN DIEREN (1934; The Netherlands)
18 m/y	INMAN et al. (1966; Mexico)
10–30 m/y	FINKEL (1959; Peru)
24 m/y	LONG & SHARP (1964; California)
27 m/y	LETTAU & LETTAU (1969; Peru)
3–12 m/y	PAUL (1944; Poland)
5–6 m/y	GARCIA NOVO (1977; South Spain)
Parabolic and related dunes	
5.5 m/y (mean)	RANWELL (1958; Western Europe)
6.7 m/y	RANWELL (1958; West England)
7.5 m/y	HANSEN (1857; Denmark)
5–8 m/y	SCHOU & ANTONSON (1960; Denmark)
max. 10 m/y	MISZALSKI (1973; Poland)
3 m/y	BROTHERS (1954; New Zealand)

Tab. 2: *Migration speed of dunes (after several authors)*.

have been situated between 1330 and 1600.

2.3 Phase III

A period of stabilization followed Phase II, supported by historical evidence (GOTTSCHALK 1977), until around 1750 when renewed sand blowing started. These processes mainly affected near the coast, although vulnerable areas elsewhere witnessed more continuous large-scale sand blowing (KLIJN 1981). Around 1850 a large-scale stabilization by man completed in the beginning of the 20th century caused an almost entire fixation of the Dutch dune area.

Summarizing the data mentioned above, Phase I must have commenced between AD 800 and 1000, Phase II before or around 1300, ending around 1600, and Phase III around 1750, ending AD 1850/1900.

3 Causes of dune formation: several hypotheses

The following causes of coastal dune formation are frequently mentioned in literature:

- **coastal accretion** favours primary dune building processes on beach or beach plains due to abundant sand transport towards the coast. This mechanism is undisputed, but not elaborated as the dunes dealt with here belong to secondary dune formations.

- **Secondary dune forming processes behind the foredune**

Secondary dune forming is triggered by destruction of plant cover and subsequent wind erosion. It includes deformation and displacement of existing dunes by deflation and/or accumulation. Pedological, biotic, antropic or climatic factors can seldom be considered to be the sole cause. A combination of factors appears to be more appropriate.

Soil forming processes, such as leaching of primarily poor sands, lead to extreme vulnerable stages in vegetation succession, such as lichens. From both geomorphological and historical evidence KLIJN (1981) concluded that certain areas poor in lime and other minerals suffered repeated sand blowing on a large scale.

Biotic influences are known to influence the conditions of plant cover. Grazing and burrowing by rabbits can locally lead to sand blowing.

Antropic influences can lead directly or indirectly to destruction of plant cover. Forest felling or cutting of shrubs, and even marram grass for fuel, overgrazing by cattle and digging in order to catch rabbits are well known activities. Sand blowing due to overexploitation by man is often mentioned in historical sources. The political or socio-economic context proved to be of significance (VAN DIEREN 1934, JELLES 1968, CHRISTIANSEN et al., this volume).

Direct climatic influence on the plant cover has hardly been studied. Long periods of drought in summer could possibly have caused serious damage, particularly when combined with overgrazing by cattle and rabbits, or by fire. Periods with extreme drought are described in historical records (LABRIJN 1945) for the years around 1800 or deduced from indirect evidence from other sandy areas for the 10th century (HEIDINGA 1984).

3.1 Erosion of the foredune after reaching equilibrium height

According to RANWELL (1972) the vertical growth of foredunes is impeded by high wind velocities. As vertical wind gradients are exponential, trapping of sand by pioneer vegetation is not possible above a critical height. Each building phase should then be followed by an erosional phase starting at the upper part of the foredunes. In actual fact this is not a very widespread phenomenon, partly because foredunes are generally very well taken care of by dune managers. Moreover, in case of ample sand sources (accretion coasts), one would expect sand accumulation seawards of the outer dune foot.

3.2 Formation of dune cliffs due to marine erosion

Intermittent marine erosion during storm surges or a more steady coastal retreat causes a steep and bare dune cliff. The unstable sand surface of this cliff is practically unprotected against strong winds, and all is the more where lack of sand along erosional coasts prevents restoration of pioneer vegetation dependent on sand deposition. This phenomenon is common along many dune coasts and must have been widespread in times when human interference in the outer dunes was minimal. The relationship between coastal erosion along coastal inlets between the wadden islands and the subsequent parabolization of the attacked foredunes was studied by VAN DIEREN (1934). The same mechanism was assumed by KLIJN (1981) for most Younger Dunes in the Netherlands.

When confronting these hypotheses relating to the Younger Dunes in the Netherlands it must be stressed that the majority of datings applies to secondary dunes that have migrated from the former coastline landinward. These dunes show a large scale zonation and a morphology that is rather uniform over large distances more or less independent from differences in soil conditions, vegetation and management history (KLIJN 1981). This means that the origin of dune formation should be related to processes in or near the foredune. As this coastline has been subject to periodic erosion during the last millennium (see par. 4), secondary dune forming had most probably been triggered by periodic dune cliff forming, followed by parabolization. Since the formation of the Younger Dunes shows somewhat distinct phases, it would be tempting to relate these to changes in climate and coastal processes.

4 Secondary dune forming in relation to climatic history and coastal development

As already indicated coastal erosion could be related to the following mechanisms:

- sea level rise or an acceleration in sea level rise

- increased storm frequency, accompanied by storm surges (KLIJN 1981, KLIJN, this volume)

Of course other factors may also influence coastal erosion (BIRD 1985), but as far as the coastline and the period involved are concerned, the above mentioned factors seem to be the most relevant. Climatic changes can be considered directly or indirectly to be the ultimate motor behind both processes.

4.1 Climatic history

Climatic history for the Middle Ages and later centuries has been studied by several authors. LAMB (1977) compiled most of the previous work in addition to his own. Tab. 3 shows the main characteristics.

In the Netherlands, the 10th century must have been dry (HEIDENGA 1984), which is in accordance with very few river floods reported by GOTTSCHALK (1971, 1975, 1977). LAMB (1977) mentions a dry 10th century for Western Europe. Transitional periods before the Little Optimum, between the Little Optimum and the Little Ice Age and thereafter are considered to have been strong in circulation and rather wet. Furthermore this is supported by higher river discharges (GOTTSCHALK 1971, 1975, 1977).

4.2 Sea level

Data on sea level rise or fall along the Dutch coast are based on measurements from AD 1682 onwards (VAN VEEN 1954) or indirect data from geological or archeological studies. VAN VEEN presented a curve (fig. 1) which shows a gradual rise from 1700 onwards until 1780, then a slight fall coinciding with a dry and cold period mentioned by LABRIJN (1945) and a rather sharp rise of approximately 20 cm per century following 1830.

Sea level rise as reconstructed for the last two millenia by BENNEMA (1954) and JELGERSMA (1961) should not have exceeded 1 meter, which amounted to 5 cm per century. LOUWE KOYMANS (1974), studying archeology in the southwestern part of Netherland, showed that with high water level rise, several periods with distinct acceleration or slowdown could be distinguished. Analogous conclusions with respect to mean sea level were drawn for other regions in Europe (MÖRNER 1969,

Period (yrs AD)	Characteristics
300–400/500	Relatively warm and dry
400/500–800	Relatively cold and wet
800–950/1150	Primarily cold, later milder winters, relatively wet summers. Possibly ending around 1150 A.D. in Western Europe.
950/1150–1200/1300	"The Little Optimum": Higher annual temperature. Warmest period in Western Europe: 1150-1300 A.D. Weaker circulation with westerly main circulation. Depression paths 3 to 4° north of current depression paths.
1200/1300–1550	Deteriorating of climate in transient period. Periodically more severe winters, mostly mild and wet. Wetter summers, increased circulation and depression activity.
1550–1700*	"Little Ice Age": relatively cold winters, and wet summers, weak circulation, depression zone 3–5° south of present zone. (* the Little Ice Age is often considered to have ended around 1850).
after 1700	Improving climate, transitional period, with the exception of a shorter period around 1800. Increase in frequency and intensity of westerly winds.

Tab. 3: *Climatic characteristics after ca. 300 AD (after LAMB 1977).*

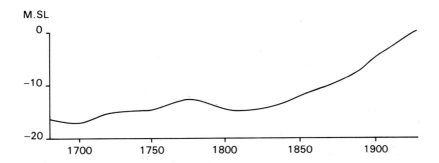

Fig. 1: *Sea level rise since app. 1682 (after VAN VEEN 1954).*

ROHDE 1978, TERS 1973, TOOLEY 1978). TOOLEY mentions a period of a falling sea level for a geologically stable area along the northwestern coast of England from approx. AD 200 to 700, followed by a rise, the maximum being around 830 to 805 B.P. (ca. AD 1150). Afterwards the sea level fell again, approximately 0.8 m in 5 centuries. Assuming that these data are valid for eustatic sea level fluctuations on other European coasts as well (see also ROHDE 1978), they should be correlated to data on isostatic or tectonic downwarp relevant for the Dutch coast. Altogether the latter mechanisms could be held responsible for 5 cm per century sea level rise at the maximum. It could be concluded that combined eustatic, isostatic and tectonic causes along the Dutch coasts must have resulted in a temporary fall (in a order of magnitude of max. 10 cm per century) as well as periods of rapid rise, the latter comparable with that at present (20 cm per century, VAN MALDE 1987).

4.3 Frequency of storms and storm surges; transgressive periods

Coastal erosion or accretion is also a function of storm surges (heights and frequencies). Unambiguous data on storm frequencies, storm floods and their effects on coasts are seldom available. Therefore indirect evidence must be used from geological studies indicating phases of transgression or regression. These phenomena are also related to sea level changes. Geological data referring to Dutch coastal areas and adjacent northwestern countries are gathered in fig. 2. For the period A.D. most authors mention three transgressive periods (Dunkerque D II, D IIIa and D IIIb). The onset of D IIIa is thought to be around AD 800, of D IIIb around 1300 A.D. However, ZAGWIJN & VAN STAALDUINEN (1975) do not distinguish Dunkerque III a and b. Remarkably, the transgressive phases are also climatologically comparable periods (transitional, increased atlantic character). The Dunkerque II and Dunkerque IIIa took place in a period of increased sea level rise and the Dunkerque IIIb in a period of decreasing rise or even a falling level, but in a concurrently stormy period. Sometimes authors distinguish phase IIIc, after the Little Ice Age (i.e. after 1700). All transgressive periods could be of major or lesser importance: Dunkerque IIIa and IIIc because of the combination of two mechanisms (increased storminess and an absolute rise in sea level), Dunkerque IIIb due to a well documented increase in storminess and storm flood frequency in the 12th and 13th century following (!) a period of increased sea level rise during the Little Climatic Optimum, which could have made coastal systems extra vulnerable to storm floods. The period after about AD 1600 witnessed both a drastic decrease in storm (flood) frequency and a decrease in sea level rise or even a temporary sea level fall until around 1700/1750 (resp. GOTTSCHALK 1975, 1977 and VAN VEEN 1954). After AD 1830 both an acceleration in sea level rise (VAN VEEN 1954) and increased storm frequencies are reported.

5 Conclusions on the relationship between climate, hydrography and dune formation

The phases of secondary dune formation, as elaborated in section 2, the climatological periods of section 4 and the peri-

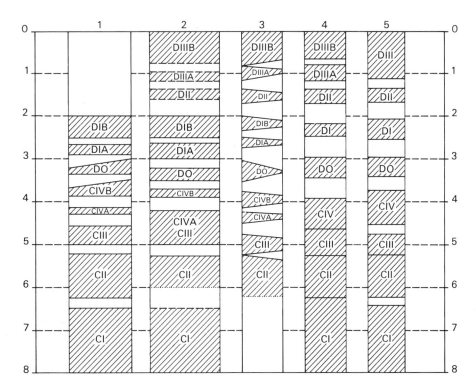

Fig. 2: *Transgressions and regressions in the Netherlands according to various authors (1 = GRIEDE 1978, 2 = ROELEVELD 1974, 3 = LOUWE KOOIJMANS 1974, 4 = VAN RUMMELEN 1972, 5 = HAGEMAN 1969; C = CALAIS; D = DUNKERQUE).*

ods of increased coastal erosion are combined in fig. 3. The following compound hypothesis on origin and chronology of the Dutch coastal dunes along the mainland coast can be formulated. Coastal dunes along the Dutch coast chiefly originate from secondary dune formation triggered by coastal erosion. Pedological, animal or human influences cannot be regarded as primary causes, except on a local scale. Frequency and severity of storm floods can be considered to be the major factor for erosion, in addition to a rising sea level. Periods of intensified coastal erosion coincide with climatologically transitional periods with increased gale frequency commonly belonging to more atlantic weather types. The effect on coastal erosion must have increased by a preceding or synchronous sea level rise. This hypothesis is summarized in fig. 4. The hypothesis mentioned above certainly diverges from the view of JELGERSMA et al. (1970) or BERENDSEN & ZAGWIJN (1984). These authors suppose an increased coastal erosion causing a steepening of the coastal profile in the submarine part. This material would be deposited in subaerial positions, thus leading to dune formation. Comparable assumptions are given by other authors (LAMB 1977, TOOLEY

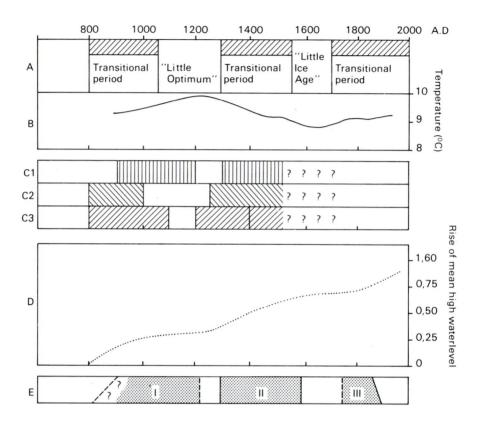

Fig. 3: *Relation of climatic history with coastal development and secondary dune formation since 800 A.D..*
A = climatic history after LAMB (1977)
B = Mean annual temperature (free after LAMB 1977)
C = Transgressive phases (after 1: VAN RUMMELEN 1972, 2: ROELEVELD 1974, 3: ZAGWIJN & VAN STAALDUINEN 1975)
D = Rise of M.H.W.L (after LOUWE KOOIJMANS 1974)
E = Phases in Younger Dune formation (KLIJN 1981)

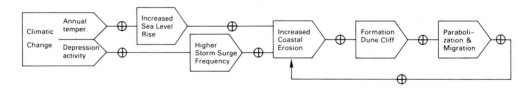

Fig. 4: *Cause-effect relationships between climate, hydrography and dune formation. Explanation in text.*

1978, and also by CHRISTIANSEN et al., this volume). In the case of coastal dunes of the Dutch mainland the last mentioned explanation is not satisfactory, because there should be indications for a rather important coastal accretion, whereas most sources point to erosion. This is true for all phases of dune formation from ca. AD 900 onwards.

Acknowledgements

The author thanks Theo Bakker and Pim Jungerius for useful comments, B. ten Cate and Mrs. De Sylva for their textural improvements and G. van Dorland for preparing the drawings.

References

BENNEMA, J. (1954): Bodem — en zeespiegelbewegingen in het Nederlandse kustgebied. Thesis, Wageningen.

BERENDSEN, H.J.A. & ZAGWIJN, W.H. (1984): Some conclusions reached at the symposium on geological change in the western Netherlands during the period 1000–1300 A.D. Geol. Mijnb. **63**, 225–229.

BIRD, E.C.F. (1985): The study of coastline changes. Zeitschr. f. Geomorph. Suppl. Band **57**, 1–9.

BROTHERS, R.N. (1954): A physiographic study of recent sand dunes in the Auckland West Coast. New Zealand Geography **10**, 47–59.

CHRISTIANSEN, C., BOWMAN, D., DALSGAARD, K. & MØLLER, J.T. (1990): Coastal dunes in Denmark: formation and present situation. This volume.

DE MULDER, E.F.J. (1983): De bodem van 's Gravenhage. Med. R.G.D., vol. **37-1**, 159 pp.

DIEREN, J.W. VAN (1934): Organogene Dünenbildung. Thesis, Den Haag.

FINKEL, H.J. (1959): The barchans of South Peru. J. Geology **67**, 614–847.

GARCIA NOVA, F. (1977): The ecology of vegetation of the dunes in Doñana National Park (South-West Spain). In: Jefferies, R.L. & Davy, A.J. (1977), Ecological processes in coastal environments. Blackwell Sc. Publ., 571–592.

GOTTSCHALK, M.K.E. (1971, 1975, 1977): Stormvloeden en rivieroverstromingen in Nederland. Deel I: De periode vóór 1400; Deel II: de periode 1400–1600; Deel III: de periode 1600–1700. Assen.

GRIEDE, J.W. (1978): Het ontstaan van Frieslands Noordhoek. Diss. Amsterdam, 186 pp.

HAGEMAN, B.P. (1969): Development of the Western part of the Netherlands during the Holocene. Geol. & Mijnbouw **48(4)**: 373–388.

HANSEN, V. (1957): Sandflugten i Thy. Geografisk Tidsskrift **56**, 69–92.

HEIDENGA, H.A. (1984): Indications of severe drought during the 10th century from an inland dune area in the central Netherlands. Geol. & Mijnbouw **63**, 241–248.

INMAN, D.L., EWING, G.C. & CORLISS, J.B. (1966): Coastal sand dunes of Guerrero Negro, Baja california, Mexico. Geol. Soc. Am. Bull. **77** (), 787–802.

JELGERSMA, S. (1961): Holocene sea level changes in the Netherlands. Thesis, Leiden.

JELGERSMA, S., DE JONG, J.D., ZAGWIJN, W.H. & VAN REGTEREN ALTENA, J.F. (1970): The coastal dunes of the western Netherlands; geology, vegetational history and archeology. Med. Rijks. Geol. Dienst, N.S. nr. **21**.

JELLES, J.G.G. (1968): Geschiedenis van beheer en gebruik van het Noordhollands Duinreservaat. Med. ITBON, Nr. 87, Arnhem.

JONG, J.D. DE (1984): Age and vegetational history of the coastal dunes in the Frisian islands, The Netherlands. Geol. & Mijnbouw **63**, 269–275.

KLIJN, J.A. (1981): Nederlandse kustduinen; geomorfologie en bodems. Thesis, Wageningen, 188 pp.

KLIJN, J.A. (1990): Dune forming factors in a geographical context. This volume.

LABRIJN, A. (1945): Het klimaat van Nederland geburende de laatste twee en een halve eeuw. Med. & Verh. KNMI nr. **102**, 49. 's Gravenhage.

LAMB, H.H. (1977): Climate, present, past & future. Vol. II. Methuen, London.

LETTAU, K. & LETTAU. H. (1969): Bulk transport of sand by the barchans of the Pampa de la Joya in southern Peru. Geomorphol. **13(2)**, 182–195.

LONG, J.T. & SHARP, R.P. (1964): Barchan dune movement in Imperial Valley, California. Geol. Soc. Am. bult. **75(2)**, 149–156.

LOUWE KOOYMANS, L.P. (1974): The Rhine-Maas Delta: Four studies on its prehistoric occupation and Holocene geology. Thesis, Leiden, University Press, Analecta Prehistorica, Leiden, nr. 7.

MALDE, J. VAN (1987): Relative sea rise of mean sea levels in The Netherlands in recent times. European Workshop on interrelated bioclimatic and land use changes, Noordwijkerhout.

MISZALSKI, J. (1973): Present day aeolian processes on the Slovenian coastline. Warszawa, IGDAN.

MÖRNER, N.A. (1969): The Late Quaternary History of the Kattegat sea and the Swedish West Coast. Sver. Geol. Unders. Afh. Serie **C.640**, 1–487.

PAUL, K.H. (1944): Morphologie und Vegetation der Kurischen Nehrung. In: Nova Acta Leopoldina, Band B, nr. **96**, 1944.

RANWELL, D.S. (1958): Movement of vegetated sand dunes at Newborough Warren, Anglesey. Journ. of Ecology **46**, 83–100.

RANWELL, D.S. (1972): Ecology of saltmarshes and sand dunes. Chapman & Hall, London, 258 pp.

RENTENAAR, R. (1977): De Nederlandse duinen in de middeleeuwse bronnen tot omstreeks 1300. Geogr. Tijdschr. **XI**, 5.

ROELEVELD, W. (1974): The Groningen coastal area. A study in Holocene geology and low-land physical geography. Ber. Rijksd. Oudh. k. Bodemonderzoek. Vol. **20**, 7–25, Vol. **24**, 7–132.

ROEP, Th.B. (1984): Progradation, erosion and changing coastal gradient in the coastal barrier deposits of the Western Netherlands. Geol. & Mijnbouw **63**, 249–258.

ROHDE, H. (1977): Sturmfluthöhen und sekulären Wasserstandanstieg an der deutschen Nordseeküste. Küste **30**, 52–143, Heide BRD.

VAN RUMMELEN, F.F.F.E. (1972): Toelichtingen bij de Geologische Kaart van Nederland (1:50.000), Blad Walcheren. Rijks Geologische Dienst, Haarlem.

SCHOU, A. & ANTONSON, K. (1960): Denmark. In: Sømme, A. (ed.), A geography of Norden. Kopenhagen.

STAALDUINEN, C.J. VAN (1977): Geologisch onderzoek van het Nederlandse waddengebied. Rijks Geologische Dienst, Haarlem.

VAN STRAATEN, L.M.J.U. (1965): Coastal barrier deposits in South- and North-Holland. Med. Geol. Stichting, Nwe Serie **17**, 41–75.

STUIVER, M. (1982): A high precision calibration of the AD radiocarbon time scale. Radiocarbon **24**, 1–26.

SUESS, H.E. (1969): Die Eichung der Radiocarbonuhr. Bild der Wissenschaft. Heft **2**.

TERS, M. (1973): Les variations du niveau marin depuis 10.000 ans le long atlantique francais. Le Quaternaire: Géodynamique, Stratigraphie et Environment National de la Recherche Scientifique, INQUA.

TOOLEY, M.J. (1978): Sea level changes; North West England during the Flandrian Stage. Clarendon Press, Oxford.

VEEN, J. VAN (1954): Tide gauges, subsidence gauges and flood stones in the Netherlands. Geol. & Mijnb. **16**, 214–219.

ZAGWIJN, W.H. (1984): The formation of the Younger Dunes on the west coast of the Netherlands (A.D. 1000–1600). Geol. & Mijnb. **63**, 259–268.

ZAGWIJN, W.H. & VAN STAALDUINEN, C.J. (1975): Toelichting bij de geologische overzichtskaart van Nederland. Rijks geologische Dienst, Haarlem.

Address of author:
J.A. Klijn
The Winand Staring Centre for Integrated Land, Soil and Water Research
P.O. Box 125
6700 AC Wageningen
The Netherlands

CHRONOLOGY OF COASTAL DUNES IN THE SOUTH-WEST OF FRANCE

C. **Bressolier**, Montrouge
J.-M. **Froidefond**, Talence
Y.-F. **Thomas**, Montrouge

Summary

Coastal dunes at the south of the Gironde estuary (SW-France) prove to consist of four systems developed in distinct periods under different climatic conditions. Datings are derived from paleosols. The first stage (3000–1000 BP) resulted in parabolic dunes, the second (2400–500 BP but also later) in barchan dunes, the last stage (since the end of the 19th century) in the present foredunes (Dune du Pilat).

Dune development seems to relate to coastal changes, whereas landward transport of dunes had a strong influence on the drainage of the hinterland.

1 French Atlantic coastal dunes

Most of the Atlantic dunes (fig. 1) were built up during the Quaternary period, when the winds were stronger than at present and were dominated by Westerlies. On the Flandres plain, dune masses are arranged in an arrow strip. Further south, on the Picardy coast, they are more extensive and until 3.5 km wide and 35 m high. In the Baie de Seine, along the Normandy coast and on the Cotentin Peninsula, dunes are littoral ridges extending between headlands. From Mont Saint Michel Bay to the Loire, they are smaller and develop at bay margins. They are composed of erosion products, containing various percentages of calcareous and organic particles. South of the Loire, dunes are low (Fromentine, Grouin du Cou), although further south the large dune system are similar to the dunes bordering the Landes coast. The bulk of the sands are of local provenance.

2 Dune system geomorphology of the French SW coast

The "Landes de Gascogne" area is underlain by a formation called "Sable des Landes" which developed from -20000 to -10000 BC, under very cold conditions (LEGIGAN 1979). It represents the most important dune system in France and extends from the Gironde estuary to Bayonne (fig. 2).

Four geomorphological dune types may be distinguished, each one forming during very different climatic periods (PENIN 1980).

1 **The parabolic dunes** formed during a regressive phase of sea level

ISSN 0722-0723
ISBN 3-923381-23-9
©1990 by CATENA VERLAG,
D–3302 Cremlingen-Destedt, W. Germany
3-923381-23-9/90/5011851/US$ 2.00 + 0.25

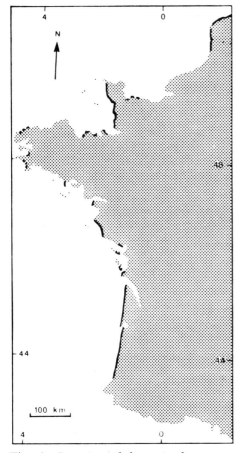

Fig. 1: *Location of the main dune systems of the French Atlantic coast.*

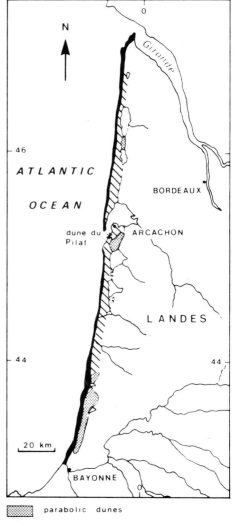

▨ parabolic dunes
■ bordering dunes and non differenciated
▨ crescentic dunes

Fig. 2: *Littoral dune types on the Aquitaine coast.*

and a relatively humid period from 3000 to 1000 BP. This climate was favourable to vegetation growth and to dune stabilization. Parabolic fixed dunes contributed to the partial or complete closure of small estuary mouths. Discharge began to be impeded behind the dune barrier forming lagoons (MANAUD 1971, LEGIGAN 1979). The dunes are U or V shaped, the windward slope is steep and faces West. Prabolic dunes range in size between 200 m and 1 km long and 10 to 40 m high. Often they join together in rake (or comb) dunes.

2   The first generation of **crescentic dunes** (or barchans) began to form during a regressive phase extending from the Iron Age (2400 BP) up to

the Gallo-Roman Period (1500 BP). Some lowering of the water table and the relatively arid climate contributed to their development. This new system completed the enclosure of the Landes lagoons: sea outlets were maintained to only two barrier lagoons (Aureilhan and Leon) up to the recent times.

Several significant increases in wind action occured and contributed to the development of other crescentic dunes: around 600 and 1300 AD (1300 respectively 650 BP) in the Middle Ages, and also during the 17th and the 18th centuries. These very mobile dunes became dangerous for human settlements and made land reclamation necessary. These reclamations started at the end of the 18th century and took the form of afforestation by pines and soil drainage and were finished in the mid 19th century (SARGOS 1949).

Crescentic dunes are set up in quasi-continuous strips, parallel to the coast with smooth windward slopes. They are usually mobile and separated by brackish interdune depressions, locally called "lettes" or "lèdes".

3   The present foredunes are associated with the other **wind constructed forms**, particularly embryonic-dunes behind obstacles, or chaotic dune patterns, formed since the 19th century, or some parabolic dunes where "caoudeyres" (blowouts) may develop when vegetation is sparce.

4   **The foredune** faces the sea, reaching 10 to 30 m in height, it extends 100 to 300 m landward. This system was stabilized since the early 20th century by fences and marram grass.

The different types of dunes are not distributed uniformly along the coast. From north to south: Crescentic dunes are dominant in the north; in the centre there are parabolic dunes extended between the lakes. In the far south (below courant d'Huchet) parabolic dunes occur, representing the oldest coastal dunes.

3   The "dune du Pilat": a case study

This dune is located at the entrance of the Bassin d'Arcachon. This Arcachon bay area constitutes an excellent model area for the study of the progressive formation of the lagoon originating from an estuary. This evolution started about 5000 years BP at the same time as a fast sea level rise, within a general transgressive context (MANAUD 1971, CARBONEL et al. 1987). The Leyre, which flows into the lagoon, is the most important river draining the Landes. The lagoon entrance is subject to very intense marine processes (high energy waves) and is particularly unstable. These processes control the sedimentary dynamics of the inlet, the lagoon channels and the adjacent shoreline. In front of the Pilat dune, the shoreline has retreated 2 km since 1708 (according to a map by Claude Masse) destroying an important dune complex and a small pond ("bassin du Pilat"). Entrance channel shifting is the main factor explaining the erosion during and since the 18th century.

Presently, the Pilat dune is an asymmetrical transversal dune, with a steep east facing slope. It is 2.5 km long, 500 m wide and 100 m high. The East side is precipitous and overlies older dunes

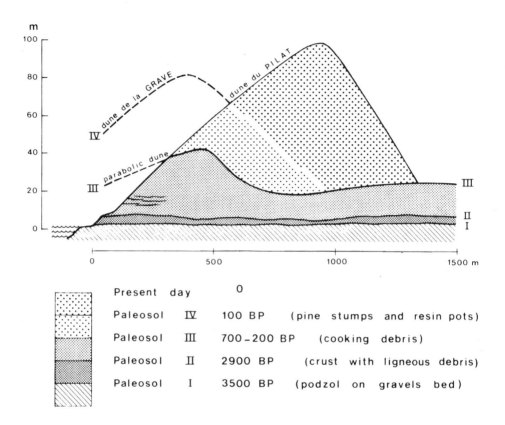

Fig. 3: *Littoral dune types on the Aquitaine coast.*

(parabolic and some crescentic dunes). On the windward side (seaward), one can observe numerous darker and undulating beds corresponding to organic deposits and the heavy mineral accumulations (a thousand or more of such strata can be observed). Six to eight leels, containing organic matter are laterally persistent. They correspond to paleosols. Only the oldest, Paleosol I, podzolised.

1 Description and datings

(FROIDEFOND 1985, PAQUE-REAU & PRENANT 1961, DAU-TANT et al. 1983) (fig. 3)

At the foot is paleosol I, at a level varying between +1 and +2 m above low tide. It is a podzol with a rather thick "alios" bed (very hard soil to be assimilated to lateritic soils) with stumps and root fragments. It overlies a coarse sand and gravel unit (PAQUE-REAU & PRENANT 1961), dated to 3680 ± 100 BP and paleosol II occurs between +2 and +4 m above low tide. It is a weakly expressed soil with a non-permeable humus layer containing ligneous debris. It overlies a grey sand unit, 10 cm thick, and corresponds to the level of upwelling groundwater causing a ferrugineous crust on top of the humus layer. FROIDEFOND (1985) observed in 1978 an outcrop of green clay below

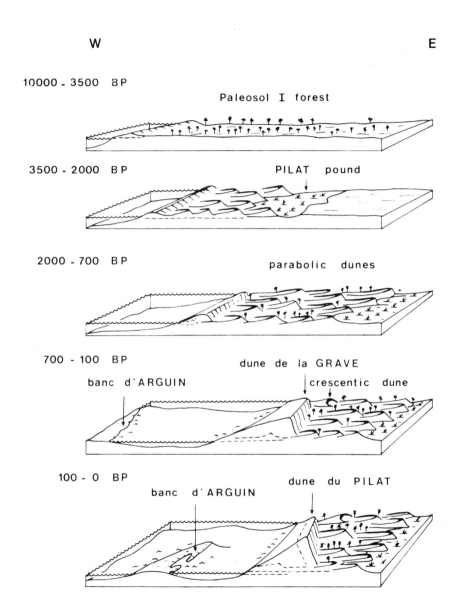

Fig. 4: *Paleogeographic reconstruction of dune system evolution and formation of the "dune du Pilat".*

this paleosol, which has been dated to 2980±100 BP.

Two or three intermediate paleosols are then observed between +10 and +15 m above the high tide mark. They are 2 to 5 cm thick, grey coloured and lack an organic horizon. They are overlain by freshwater diatoms valves (FROIDEFOND & LEGIGAN 1985). This bed could have formed under selective wind action from the lacustrine environment located further eastward. Charcoal fragments (prehistorical location at +8.5 m) gave a date of 2690±100 BP (DAUTAND et al. 1983).

Paleosol III is at +20–40 m. An organic horizon, 5 to 30 cm thick, overlies 10 to 20 cm of grey sands (paleosol with cooking debris, shell fragments, shards and chipped flint implements). This level is very undulating and is dated, from pottery typology, to between the 13th and 17th century. Ir corresponds to the parabolic dune soil (FROIDEFOND et al. 1983).

Paleosol IV crops out at about +60 m. A level of grey sands, 10 to 20 cm thick is overlain by an organic horizon. Numerous pine stumps and resin pots can be observed. On the southern part of the dune, this level truncates paleosol III at +5 m. As this last level developed on the dune surface, paleosol IV is dated to 1850–1890 ("Conseil Général" map).

2 Paleogeographical reconstruction (fig. 4)

From -10000 to -3500 BP, a large forest with pine, birch wood and hazel trees covered a sandy soil (PAQUEREAU & PRENANT 1961). The sea was 6 to 8 km further west. Around -3500 BP. a dry period began: strong winds brought sands to bury this forest. New vegetation (paleosol II) began to colonize these weakly developed dunes. Around 2500 BP, the sea was eroding these dunes and wind blown sands began to choke the river mouths, forming lagoons which were later covered by parabolic dunes. Towards 2000 BP, man settled in the sparse pine/hazel forest. Around 1500 BP, climatic change and/or sea-level rise disturbed the equilibrium. Crescentic dunes, which only occurred along the shoreline, invaded the parabolic dune system and buried paleosol III below 50 m of sand. One of them is the "dune de la Grave". These dunes moved very quickly. They have been stabilized by artificial means since the end of the 18th century.

In 1863, the "dune de la Grave", was located at the East of "Banc d'Arguin" (from the Conseil Général map). When the main entrance channel migrated to the foot of the dune, vegetation began to be stripped away, and the sands were eroded. These sands nourished the Bernet bank and the inlet bar, but wind action increased the eatward progression of the dune and resulted in the present Pilat dune construction.

One or several of the stages observed at the Pilat dune correspond to the major type of dune settlements on the French Atlantic coasts. The Pilat dune has stratigraphically recorded traces of the major part of French Atlantic dune construction.

References

BRESSOLIER, C. (1984): Bibliographie analytique des Côtes de France, Landes et Pays Basque. Mémoires du Laboratoire de Géomorphologie. EPHE 37, 197.

COSTE, M. (1978): Inventaire succinct des Diatomées recueillies dans les sables de la dune du Pilat. Personnal communication, in FROIDEFOND (1985).

DAUTANT, A., JACQUES, P., LESCA-SEIGNE, A. & SEIGNE, J. (1983): Découvertes protohistoriques récentes près d'Arcachon. Bulletin Société française des Pétroles **80**, 188–192.

FROIDEFOND, J.-M. (1985): Méthode de géomorphologie côtière. Application à l'étude de l'évolution de littoral Aquitain. Mémoire Institut géologique du Bassin d'Aquitaine **18**, 273, 7 pl.

FROIDEFOND, J.-M. & LEGIGAN, P. (1985): La Grande Dune du Pilat et la progression des dunes sur la littoral Aquitain. Bull. Inst. Géol. du Bassin d'Aquitaine **38**, 69–79.

LEGIGAN, P. (1979): L'élaboration de la formation du sable des Landes, dépôt résiduel de l'environnement sédimentaire Pliocène Pléistocène centre Aquitain. Mémoire Institut géologique du Bassin d'Aquitaine **9**, 429.

MANAUD, F. (1971): L'évolution morphologique récente du Bassin d'Arcachon. Thèse 3e cycle, Université Bordeaux III, 102, 42 pl.

NAVE, F. (1983): Protection des dunes contre l'érosion marine et éolienne. Journées d'information ASTEO, "Erosion et défense des côtes", Paris, 26–27 Janvier 1983, **III-4**, 30.

PAQUEREAU, M.-M. & PRENANT, A. (1961): Note préliminaire sur l'étude morphologique et palynologique de la grande dune du Pilat (Gironde). Procès-Verbaux Société linnéenne de Bordeaux **99**, 27–38.

PÈNIN, E. (1980): Le prisme littoral aquitain: Histoire Holocène et évolution récente des environnements morpho-sédimentaires. Thèse 3° cycle, Université Bordeaux I, 129.

Addresses of authors:
Catherine Bressolier
Laboratoire de géomorphologie EPHE
1 rue Maurice Arnoux
92120 Montrouge
France
Jean-Marie Froidefond
Département de Géologie et Oceanographie
URA 197 du CNRS
351 Cours de la Université de Bordeaux
Avenue des Facultes
33405 Talence
France
Yves-Francois Thomas
Lab. de Géographie Physique
URA 141 du CNRS
1 rue Maurice Arnoux
92120 Montrouge
France

JUST PUBLISHED

Aaron Yair & Simon Berkowicz (Editors)

ARID AND SEMI-ARID ENVIRONMENTS

Geomorphological and Pedological Aspects

Selected papers on the Workshop on Erosion, Transport & Deposition Processes, IGU Commission on Measurement, Theory & Application in Geomorphology, Jerusalem, Sede Boqer, Elat, March 29 - April 4, 1987

CATENA SUPPLEMENT 14

hardcover/176 pages/numerous figures, photos and tables

ISSN 0722-0723/ISBN 3-923381-17-4

list price: DM 98.-/US $ 62.-/ standing order rate CATENA SUPPLEMENTS: DM 68.60 /US $ 43.40

ORDER FORM

☐ Please send me at the rate of DM 98.-/ US $ 62.- .... copies of CATENA SUPPLEMENT 14.

☐ I want to subscribe to the CATENA SUPPLEMENTS starting with no. ....

Name ....

Address ....

Date ....

Signature: ....

Please charge my credit card: ☐ MasterCard/Eurocard/Access ☐ Visa ☐ Diners ☐ American Express

Card No.: .... Expiration date: ....

Please, send your orders to:

CATENA VERLAG, Brockenblick 8, D-3302 Cremlingen-Destedt, West Germany, tel.05306-1530, fax 05306-1560

USA/Canada:**CATENA VERLAG**, Attn. John Breithaupt, P.O.Box 368, Lawrence, KS 66044, USA, Tel. (913) 843-1234, fax (913) 843-1244

THE GEOHYDROLOGY OF COASTAL DUNES

T.W.M. Bakker, The Hague

Summary

Geohydrology is a component of the landscape which, because of the flowing character of groundwater, links landscapes together. Human activities in one area may have (severe) consequences in another. Biological and ecological features of dunes are to a large degree defined by quantity and quality of the groundwater.

A review is given of the flow of groundwater in dune areas with a high- and a low-lying hinterland. Components like climate, geology, geomorphology, vegetation and human activities which influence this flow as well as the water quality, are discussed.

1 Introduction

The fresh water of coastal dunes is of great importance for their biological and ecological features. It defines to a certain degree patterns of vegetation, geomorphology and even geology. Its role in creating the wet slack environment of dunes must especially be mentioned. Since the middle of the nineteenth century the fresh watercontent serves in some countries of Europe as a source of drinking or irrigation water for the local population.

It lasted however until around 1920 before the groundwaterflow of dune areas was understood and mathematically described. This description was very quantitative. In the last decades this has changed appreciable. The great importance of the waterquality of dunes for nature conservation was recognised. Ecological research in which quantity and quality of water played a central role, started. RANWELL (1975) in Great Britain and LONDO (1971) in the Netherlands did pioneering work. the occurrence of plant species in relation to quantity and quality of (ground)water was extensively studied. A new school of research developed: the so called ecohydrological approach.

Closely related to this is the so-called 'systems approach' of groundwater flow (see ENGELEN & JONES 1987). Stuyfzand's geochemical investigations of dunes and hinterland are a further development of this approach (see STUYFZAND, this volume).

Despite this research we must acknowledge that dune hydrology in most European countries is rather undeveloped. Basic information about the amount of precipitation, the mapping of groundwatertables and geological structure is lacking.

In this paper an outline is presented of the occurrence of fresh groundwater in

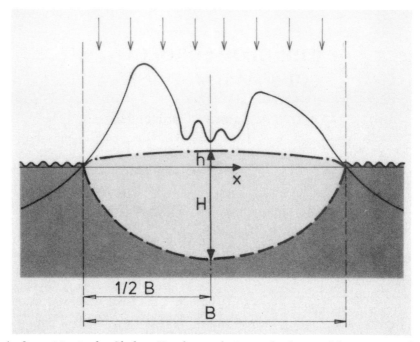

Fig. 1: *Quantities in the Ghyben-Herzberg relation and scheme of flow or groundwater for a dune area with a low-lying hinterland.*

coastal dunes and of factors influencing its flow and quality. Three main subjects are discussed:

1. Flow of groundwater
2. Factors influencing the flow, the quantity and quality of groundwater
3. Human influences on groundwater-flow

2 The flow of groundwater

2.1 The Ghyben-Herzberg relation

BADON GHYBEN in 1889 and HERZBERG in 1901 argued that the fresh and the salt water in the subsoil are in equilibrium with each other. At the fresh-salt interface the pressure of the fresh water above must equal the pressure of the salt water below (see fig. 1).

In symbols:

$$(h + H) \cdot \rho_f = H \cdot \rho_s$$
$$H = \frac{\rho_f}{\rho_s - \rho_f}$$

Meaning of symbols:
- h: height of groundwatertable above sealevel (m)
- H: depth of fresh water below sealevel (m)
- ρ_f: density of fresh water (kg/m³)
- ρ_s: density of salt water (kg/m³)

This so-called Ghyben-Herzberg relation is a simple mathematical description of a very important geohydrological phenomenon in coastal dunes: the fresh groundwater floats on the salt groundwater. Knowing that $\rho_f = 1000$ kg/m³ and $\rho_s = 1025$ kg/m³, the depth of the body of fresh water below sealevel should be forty times the height of this body above

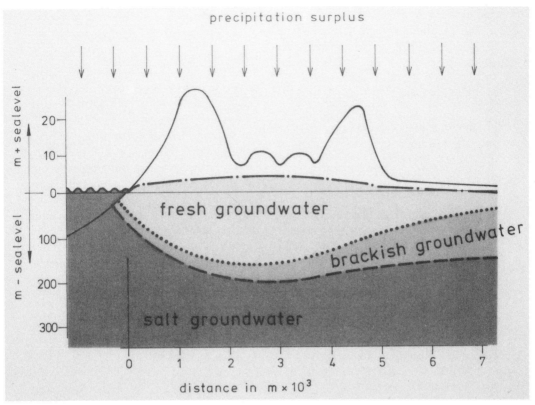

Fig. 2: *Cross-section of a coastal dune area and adjoining land with schematic flow of groundwater.*

sealevel. In reality, however, this value is never reached. Values of fifteen to twenty are normal.

The flow of groundwater in coastal dunes is much more complicated than is suggested by the scheme of Ghyben-Herzberg. Fig. 2 shows a more realistic scheme in which also the role of precipitation as the driving force behind the dynamic equilibrium between the fresh and the salt groundwater, is demonstrated.

In the following this scheme, with its more dynamic view of the geohydrology of dune areas, is worked out. Three situations, which are common along the European coast, are distinguished.

1. Dunes with a low-lying hinterland in humid areas
2. Dunes with a high-lying hinterland in humid areas
3. Dunes in arid areas

2.2 Dunes with a low-lying hinterland in humid areas

Dune areas with a low lying hinterland, see fig. 2, are quite common in the lowlands of Poland, Germany, Denmark, The Netherlands and Belgium. Because of the more or less permanent input of water by precipitation a flow of groundwater can be maintained. The water leaves the dune area by means of underground flow to the sea on one hand

and to the inner dune fringe on the other. This situation can be described in simple mathematical terms. Darcy's law yields:

$$Q = -k \cdot (H + h) \cdot \frac{dh}{dx}$$

The principle of continuity, together with the condition that $Q = 0$ when $x = 0$, requires:

$$Q = N \cdot x$$

Together with the Ghyben-Herzberg relation, the description of the height of the groundwatertable in the dune area is as follows:

$$h^2 = (\delta \cdot N/k) \cdot (B^2/4 - x^2)$$

Meaning of symbols:
N: Precipitation surplus (Precipitation minus evapotranspiration) (m/day)
k: Specific conductivity of subsoil (m/d)
δ: $(\rho_f)^2 / \{\rho_s \cdot (\rho_s - \rho_f)\}$
B: Width of dune-area (m)
x: Distance from middle of dune area (m)

From this expression we can learn that the groundwatertable (h) is dependent on the amount of the precipitation surplus (N), the permeability of the subsoil (k), the width of the dune area (B) and on the distance from the middle of the dune area (x). The permeability, in turn, depends on the lithology, and especially on the grain size of the dune sands.

During the travel of the groundwater through the subsoil of the dune area, the quality of the water changes. In the beginning, certainly in dune areas poor in lime, the groundwater is acid and resembles precipitation water. During its flow, soil components dissolve and the waterquality changes. STUYFZAND (this volume) discusses these changes and describes patterns of groundwater quality in the subsoil.

2.3 Dunes with a high-lying hinterland in humid areas

Dune areas with a high lying hinterland, see fig. 3, are common in parts of Europe which have rocky coasts. So-called climbing dunes are blown more or less over these rocks. The feeding of the groundwater of these dune areas is twofold. On one hand there is the direct input by the precipitation that falls on the dunes. On the other hand there is supply by flow of groundwater from the hinterland or by rivers infiltrating into the dune sands.

Also for this situation a simple mathematical description is presented. Darcy's law, yields:

$$Q = -k \cdot (H + h) \cdot \frac{dh}{dx}$$

The principle of continuity says:

$$\frac{dQ}{dx} = N$$

Together with the Ghyben-Herzberg relation and the boundary conditions that $h = 0$ when $x = 0$ and that $h = h_r$ when $x = L$, the description of the height of the groundwatertable becomes:

$$\begin{aligned} h^2 &= -\{N/(k \cdot \delta)\} \cdot x^2 + (h_r^2 \cdot \delta/L) \\ &\quad \cdot x + \{(N \cdot L)/(k \cdot \delta)\} \cdot x \end{aligned} \quad (1)$$

Meaning of symbols:
L: Distance between shoreline and inland groundwater boundary h_r (m).
h_r: Groundwatertable at the inland boundary (m above mean sealevel)
others: See before and fig. 3.

From this discription we can learn that the groundwatertable in the dune area is dependent on the amount of the precipitation surplus (N), the permeability (k) and the boundary conditions at the inner dune fringe, e.g. L and hr.

Geohydrology of Coastal Dunes

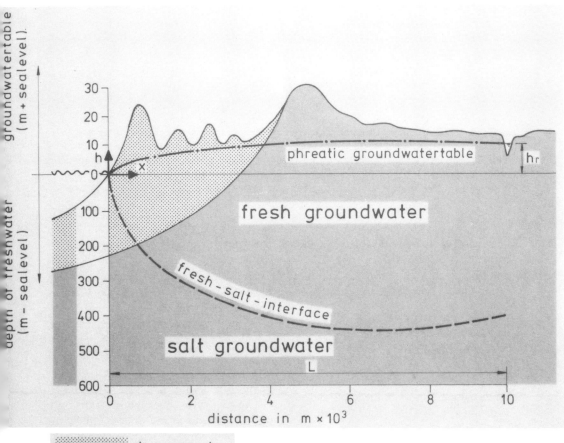

Fig. 3: *Scheme of the flow of groundwater according to the Ghyben-Herzberg relation, Darcy's law and the principle of continuity, for a dune area with a high-lying hinterland.*

Depending mainly on the size of the hinterland and the composition of the rocky subsoil, there is a flow of groundwater to the dune area. This water, originating from the hinterland, may contaminate the dune area with polluted inflows.

2.4 Dunes in arid areas

In arid areas the precipitation surplus (N) is small or may even be zero. Because of this the body of fresh water will also be small or even be absent. The upper groundwater will be considerably influenced by the salty water descended from the sea. On the other hand, when the hinterland is large enough, there may also

Type of vegetation	Evapotranspiration (mm/y)	value of f (-)	value of g (-)
Bare dunes	180	—	0.25
Wet slack vegetation	550	0.7	—
Dry dune vegetation	360	—	0.5
Wet decidious woodland	550	0.7	—
Dry decidious woodland	400	—	0.55
Wet coniferous woodland	700	0.9	—
Dry coniferous woodland	550	—	0.75

Tab. 1: *Annual evapotranspiration and values of f and g, for seven types of vegetation in the dunes of the Netherlands, under average climatic conditions.*
Precipitation = 725 mm/year; Evapotranspiration of open water (according to the formula of Penman) = 770 mm/year.

be a considerable supply of fresh water from this side. LLAMAS (this volume) describes such a situation for the dunes of the Coto Doñana in southern Spain.

3 Factors influencing the geohydrology

In the next section a review is given of some specific factors influencing the geohydrological situation of coastal dunes.

3.1 Climate

Specific information on this subject is not presented. Here we deal with two items, which are not really a part of the climate, but of great importance to the hydrology of dunes:

1. the so-called sea-spray mechanism
2. The role of vegetation in evapotranspiration.

3.1.1 Sea-spray

Because of the wind and the waves small particles of seawater are thrown into the air and transported to the dunes. There is a rapid decrease of the saltload from this sea-spray, from 200 kg Cl/ha · year directly behind the coastline to 60 kg Cl/ha · year at two kilometres inland. After this the load slowly drops to 40 kg Cl/ha · year at 50 kilometres from the coast. Because of this the concentration of chloride in the groundwater varies from some tens of mg/l at the inner dune fringe, to a few hundreds of mg/l at places directly behind the coastal line.

The sea-spray mechanism not only brings chloride to the dunes. Other important ions are: K, Na, Mg and SO_4.

3.1.2 Vegetation

Although the amount of evapotranspiration of a dune area is in the first place determined by climatic circumstances, the type and the amount of vegetation is also of great importance. BAKKER (1981) studied this subject in the dunes of the Netherlands. Tab.1 gives a summary of his results.

Two types of vegetation are distinguished: the first growing on wet places under the influence of groundwater and the second growing in dry places, using soil moisture for their growth. Assumed is that the evapotranspiration of the first

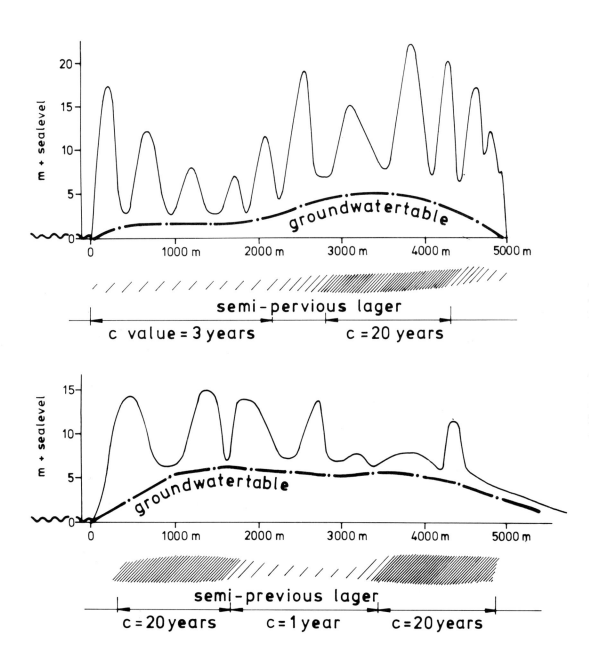

Fig. 4: *Two cross-sections, at right angles to the coast, of the Kennemerdunes near Haarlem in the Netherlands, showing the influence of semi-impervious layers on the groundwatertable. (After DUBOIS 1909).*

type of vegetation is dominated by the amount of energy that is available from the atmosphere. Thus, it is strongly correlated with the amount of evapotranspiration of open water. In symbols:

$$E_a = f \cdot E_0$$

Meaning of symbols:
E_a: Actual evapotranspiration (mm/year)
f: Coefficient derived from literature (See for instance BAKKER 1981)
E_0: Evapotranspiration of open water according to Penman (mm/year)

The evapotranspiration of the second type of vegetation is assumed to be strongly dependent on the amount and temporal distribution of precipitation. In periods with sufficient precipitation, evapotranspiration may be high. In dry periods on the contrary, potential evapotranspiration will be high, but the real evapotranspiration of such vegetations may be small, because of withering or active reduction of the transpiration by the plants. For this type of vegetation the following relation between evapotranspiration and precipitation is assumed:

$$E_a = g \cdot P$$

E_a: See above
g: Coefficient derived from literature (see for instance BAKKER 1981)
P: Total amount of precipitation (mm/y)

3.2 Impervious layers

In the subsoil of dune areas semi-impervious layers of peat and clay are frequently encountered. Such layers have a considerable influence on the flow of groundwater. Fig. 4 gives two cross-sections of a dune area in the Netherlands with a subsoil of varying lithology. Where semi-impervious layers occur the groundwatertable is higher than in those places where a sandy subsoil is present.

Fig. 4 also shows that there is a close relation between the levels of slacks and the groundwatertable. This is a consequence of the development of slacks by deflation. Wind erosion goes on until the wet zone resulting from capillary rise is reached, which is usually the case at about 1/2 m above the phreatic level.

3.3 Geomorphology

Fig. 5 demonstrates that the groundwatertable under ridges is slightly higher than under slacks. RANWELL (1975) observed the same phenomenon.

These small "waves" of the groundwatertable are of importance to the quality of the groundwater, especially in areas where the dune sand is poor in lime. ESSELINK et al. (1989) investigated this subject on the Frisian island of Schiermonnikoog. Fig. 6 shows the different types of groundwater which occur in a wet slack. The narrow ridges are completely nourished by precipitation water and the groundwater in those parts is acid. From the high ridges bordering the slack, alkaline groundwater, rich in minerals, flows towards the slack and interferes with this acid groundwater. Because of this the groundwater under these central parts is alkaline, whereas more acid, "mixed" waters are found near the rims of the slacks.

4 Human influences

Human influence on the geohydrology of dune-areas may be severe. In some parts of the European coastal dunes it has caused a strong decline of natural values, especially of the wet dune slack environment.

Geohydrology of Coastal Dunes

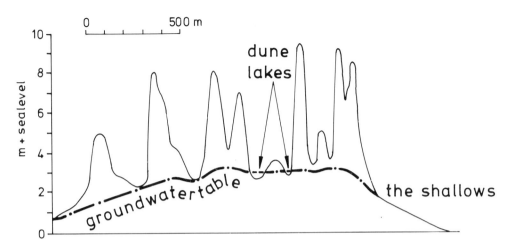

Fig. 5: *Cross-section of watertable and relief of dunes from Het Oerd, of the Frisian island of Ameland. (After BEUKEBOOM 1976).*

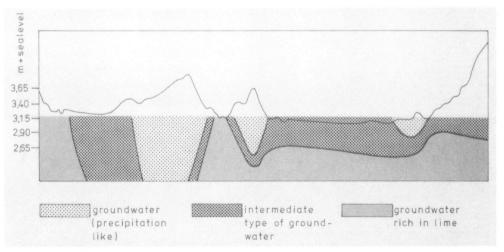

Fig. 6: *The quality of groundwater in relation to (micro)geomorphological induced patterns of groundwaterflow on Schiermonnokoog, The Netherlands. (After ESSELINK et al. 1989).*

4.1 Air pollution

Tab. 2 shows the concentration of some chemical components in the precipitation in the Netherlands, compared to values of undisturbed (?) areas. Nowadays most components show a considerable higher deposition than is to be expected in undisturbed areas. Little is known of the consequences of this for the quality of the surface- and groundwater in dune areas. VERTEGAAL et al. (1989) argue that a considerable part of the rapid increase of grassy and shrub vegetations

	"natural" conditions (VERMEULEN 1977) (kg/ha · year)	Terschelling 1984–1985 (SCHUURKENS 1987) (kg/ha · year)
Chloride	75–250	90
Sulphate	50–100	50
Sodium	40–150	50
Potassium	2– 5	9
Calcium	2– 10	10
Magnesium	5– 20	9
Ammonia	5– 15	10
Nitrate	20– 40	25
Ortho-phosphate	2.5	—

Tab. 2: *Amount of supplied components per unit of time and per unit of groundarea in "natural" circumstances according to VERMEULEN (1977) and on the Frisian island of Terschelling during the years 1984 and 1985 according to SCHUURKENS (1987).*

which is observed in the Netherlands in the last decades is due to air pollution.

Another consequence of air pollution is the so called greenhouse-effect which causes higher temperatures, a rise of the sealevel and a more rapid retreat of the coastline, as is discussed by VAN DER MEULEN (this volume).

4.2 Evapotranspiration

The data of tab. 1 demonstrate that changes in cover and type of vegetation are of great importance to the geohydrological situation of dune areas. When, for instance, bare dunes are planted with coniferous trees, the precipitation surplus may drop appreciably. From the Netherlands it is known (BAKKER 1981) that such an interference has caused a fall of the groundwatertable of 1 metre and more. Because of this wet dune slacks have dried and lost their very vulnerable vegetation.

4.3 Waterlevels at the inner dune fringe

When the waterlevel of ditches or brooklets at the inner dune fringe is changed, the groundwatertable in the main dune area will change also, especially in the zone directly boarding the inner dune fringe. This is particularly serious in dunes with a low-lying hinterland which is grained for agricultural purposes.

4.4 Extraction of drinking water

Especially in dune areas with a low-lying hinterland, the fresh water under the dunes is of great importance as a source of drinking water for the people living in the neighbourhood. A dramatic fall of the groundwatertable is in almost all cases the first, easy to observe, consequence.

4.5 Artificial infiltration of surface water

The disturbance of the geohydrology and the lowering of the groundwatertable, caused by the extraction of drinking water, urged the municipal watersuppliers to look for ways to restore the body of fresh water. Infiltration of surface water, often originating from rivers, was applied. In this way the amount of fresh water in dunes was more or less brought back to its former level. The quality of the fresh water is, however, severely affected. The amount of some chemical components, such as Cl^-, SO_4^{2-}, K^+, Na^+, Mg^{2+} and NO_3^-, which are brought into the dunes are enlarged a hundredfold and more.

5 Further research

Most of the European dunes are, from a hydrological point of view, white spots on the map. Yet geohydrology is a very important factor in understanding and managing of dunes. A mapping of geohydrological features of dunes is an enormous task to be done.

Scientific tools, especially dealing with simple methods which take into account both quantity and quality of water, must be developed to support this mapping. Ecohydrological, geochemical and geohydrological modelling are useful techniques for this purpose.

References

BADON GHYBEN (1889): Nota in verband met de voorgenomen putboring nabij Amsterdam. Tijdschr. Kon. Inst. Ingenieurs.

BAKKER, T.W.M. (1981): Nederlandse kustduinen: Geohydrologie. Thesis, Pudoc Wageningen, 189 pp.

BEUKEBOOM, Th.J. (1976): The geohydrology of the Frisian islands. Thesis Amsterdam, 121 pp.

DUBOIS, E. (1909): De prise d'eau der Haarlemsche waterleiding.

ENGELEN, G.B. & JONES, G.P. (1987): Developments in the analyses of GroundwaterFlow Systems. IAHS publ. nr. **163**.

ESSELINK, H. GROOTJANS, A., HARTOG, P. & JAGER, Th. (1989): Kalkrijke vegetaties in een duinvallei op Schiermonnikoog. Duin, 1989 **2**, 75–79.

HERZBERG, (1901): Die Wasserversorgung einiger Noordseebader. J. Gas beleucht. Wasserversorg.

LONDO, G. (1971): Patroon en proces in duinvalleivegetaties langs een gegraven meer in de Kennemerduinen. RIN, Leersum, 279 pp.

RANWELL, D.S. (1975): Ecology of salt marshes and sand dunes. London, 258 pp.

SCHUURKENS, R.J.J.M. (1987): Acidification of surface waters by atmospheric deposition. Thesis, Nijmegen, 160 pp.

VERMEULEN, A.J. (1977): Immissieonderzoek met behulp van regenvangers; opzet, ervaringen en resultaten. PWS Haarlem, 109 pp.

VERTEGAAL, C., v.d. SALM, I.N.C. & JANSEN, M.P.J.M. (1989): Omvang en oorzaken van effekten van atmosferische depositie in de duinen. Buro Duin & Kust, 47 pp. + bijlagen.

Address of author:
T.W.M. Bakker
Dunewater Company of South Holland
P.O. Box 710
2501 CS The Hague
The Netherlands

NEW

SOIL EROSION MAP OF WESTERN EUROPE

prepared by

Jan de Ploey, Leuven

In collaboration with:

Dr.A-V.Auset, France, Prof.Dr.H.-R. Bork, FR Germany, Prof.Dr.N. Misopolinos, Greece, Prof.Dr.G.Rodolfi, Italy, Prof.Dr.M.Sala, Spain, Prof.Dr.N.G.Silleos, Greece

Will the "green" Europe suffer from increasing soil degradation and even from progressing desertification over the next decade? Will Europeans be able to develop a global strategy of adequate land and water use management? The map cannot give an answer to such questions, for it merely intends to assist us in global analysis of the situation and reflection on facts and causes. The map depicts the major aspects of soil erosion in western and southern Europe.

three thematic maps/a satellite map of Western and Southern Europe/accompanied by an explaining text: Losing our Land

ISBN 3-923381-20-4

list price: DM 17,50/US $ 9.80

ORDER FORM

☐ Please send me at the rate of DM 17,50/ US $ 9.80) .... copies of SOIL EROSION MAP OF WESTERN EUROPE.

Name ....

Address ....

Date ....

Signature: ....

Please charge my credit card: ☐ MasterCard/Eurocard/Access ☐ Visa ☐ Diners ☐ American Express

Card No.: .... Expiration date: ....

Please, send your orders to:

CATENA VERLAG, Brockenblick 8, D-3302 Cremlingen-Destedt, West Germany, tel.05306-1530, fax 05306-1560

USA/Canada:**CATENA VERLAG**, Attn. John Breithaupt, P.O.Box 368, Lawrence, KS 66044, USA, Tel. (913) 843-1234, fax (913) 843-1244

HYDROCHEMICAL FACIES ANALYSIS OF COASTAL DUNES AND ADJACENT LOW LANDS: THE NETHERLANDS AS AN EXAMPLE

P.J. **Stuyfzand**, Nieuwegein

Summary

A recently developed hydrochemical facies analysis is applied to a coastal dune area and adjacent low lands in the NW-Netherlands, in order to diagnose and map the major factors accounting for variations in hydrochemistry. It implies the identification of separate hydrosomes (waterbodies with a distinct origin) and the determination of the distribution of different hydrochemical facies (zones) within each hydrosome.

The following hydrosomes have frequently been identified and are discussed: coastal dune water, polder water, relict tidal flat waters, intruded North Sea water and connate Lower-Pleistocene marine waters.

The facies of these hydrosomes is described by a combination of five facies parameters: the chemical watertype, redox index, pollution index, mineral saturation indices and the temperature. The hydrochemical evolution of dune water in the direction of ground water flow, is shown to be dictated by air pollution, the neutralisation of acids by shell fragments mainly, the consumption of oxidants, cation exchange due to intrusion and finally, in its intrusion front, the mixing with relict tidal flat water.

1 Introduction

Coastal areas constitute a unique hydrochemical district, rich in variety and dynamics (STUYFZAND 1989e). Typical is a very broad spectrum in:

- the origin of ground water, which may be e.g. the sea, lagoon or tidal flats, river or rain;

- salinity, which depends a.o. on the origin, mixing, vegetation (evapotranspiration, interception of sea spray), distance to the coast and the dissolution of evaporites;

- alkalinity, being close to zero in decalcified dunes and amounting to 60 meq/L in methane rich lagoonal groundwaters in equilibrium with calcite;

- redox potential, varying from oxic in the upper dune sands to anoxic, methanogenic in lagoonal peaty clay;

- the extent of ion exchange due to fresh or salt water intrusion and acidification; and alas

- pollution, as a consequence of human activities.

The dynamics are related to coastal progradation or erosion, fast vegetational successions, periodicity of onshore winds and man.

2 Methodology of hydrochemical facies analysis

2.1 Principles

The hydrochemical facies analysis differs from classical approaches by the integration of six aspects: the origin, chemical watertype, saturation index (with respect to minerals relevant to the system), redox level, degree of pollution and temperature (STUYFZAND 1990). It consists of five successive steps: the gathering and selection of hydrochemical data (not treated here), the objective determination of the hydrochemical facies of each sample, the identification of its origin, the construction and description of maps and cross sections and finally their interpretation.

A hydrosome is defined as a water body with distinct origin (e.g. local rain water versus artificially recharged river water running from remote mountains), and generally with a common recharge area (e.g. the dunes). Within a given hydrosome the chemical character of water varies in time and space, due to changes in recharge composition and in flow patterns and due to chemical processes between water and its porous medium. Such variations in chemical character can be used to subdivide a hydrosome into characteristic zones or "hydrochemical facies", a term introduced by BACK (1960).

2.2 Defining the facies

The hydrochemical facies of a water sample is determined by combination of the chemical watertype, redox index, pollution index, mineral saturation indices and the temperature class. The calculation of the chemical watertype, redox and pollution index is treated subsequently in section 2.3 through 2.5. The saturation index with respect to many minerals relevant to the area, like calcite, dolomite, gypsum and halite, can be calculated simultaneously with the computer program WATEQF by PLUMMER et al. (1976), whereas a simple approximation can be choosen for calcite (STUYFZAND 1989a). The temperature is subdivided into six classes (<5, 15–25, 25–40, 40–80 and >80°C; STUYFZAND 1990).

The scale and complexity of the studied area determine whether several facies-parameters need association or even omission. The facies-parameters proposed in tab. 1 may serve studies on a regional scale. Aside from the chemical watertype, three facies-parameters at most should be able then to describe the facies. Their composite code is written next to the code for the hydrosome: for example "D_{ceo}" or more concisely "D" denotes calcareous, NaKMg-equilibrated, oxidized coastal Dune water, which is, by not naming them, not gypsiferous and nonpolluted. Even the omission of "ceo" is allowed (tab. 1).

2.3 Classification of water types

STUYFZAND (1986, improved in 1989c) proposed a hydrochemical classification system with many advantages for coastal areas in particular. The determination of a water type implies the successive calculation of the main type,

Hydrochemical facies		
code	name	specification
a	acid	$SI_{caco3} < -1.0$
c	calcareous*	$SI_{caco3} > -1.0$
e	NaKMg-equilibrated#	{Na+K+Mg}-equilibrium (see tab. 4)
f	NaKMg-excess (freshened)	{Na+K+Mg}-excess (see tab. 4)
g	gypsiferous	$SI_{caso4} > -0.3$
o	oxidized@	redox index ≤ 4 (see fig. 2)
p	polluted	pollution index > 2.5 (see tab. 5)
r	reduced	redox index ≥ 5 (see fig. 2)
s	NaKMg-deficit (salinized)	{Na+K+Mg}-deficit (see tab. 4)
**	watertype	see fig. 1

*, #, @ = opposite to resp. a, f+s, and r, and therefore omissible;
SI = saturation index = log {ion activity product/solubility constant};
ALK = alkalinity;
** = for code see fig. 1

Tab. 1: *Major hydrochemical facies-parameters for coastal plains, in alphabetic order (after STUYFZAND 1990). A composite code like "cfr" plus a code for the watertype should normally be sufficient to describe the facies.*

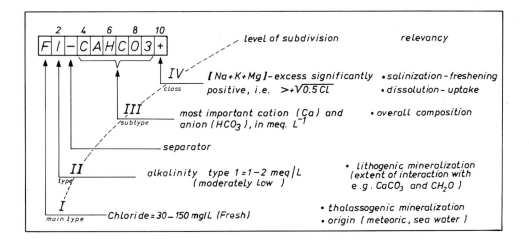

Fig. 1: *The hydrochemical classification system of water types by STUYFZAND (1989c), with its coding in 10 positions. The example is called " a fresh, moderately low alkalinity, calciumbicarbonate water with a {Na+K+Mg}surplus". This surplus is often due to a (former) intrusion of the fresh hydrosome into a saltier subsoil. A shorter notation is F_1CaHCO_3+.*

Main type	code	Cl (meq/L)	Main type	code	Cl (meq/L)
extremely fresh	G	<0.141	brackish	B	8.462-28.206
very fresh	g	0.141–0.846	brackish-salt	b	28.206-282.064
fresh	F	0.846–4.231	salt	S	282.064-564.127
fresh-brackish	f	4.231–8.462	hyperhaline	H	>564.127
Boundaries in mg/L: 5, 30, 150, 300, 1000, 10000, 20000.					

Tab. 2: *Division in main types on the basis of chloride concentration (adapted after STUYFZAND 1989c).*

type	alkalinity code meq/L	code	type	alkalinity code meq/L	code
very low	<1/2	*	very high	16- 32	5
low	1/2- 1	0	rather extreme	32- 64	6
moderately low	1- 2	1	extreme	64-128	7
moderate	2- 4	2	very extreme	128-256	8
moderately high	4- 8	3	extraord. extreme	>256	9
high	8-16	4			
Boundaries in mg HCO_3/L approx.: 30, 60, 120, 250, 500, 1000, 2000, 4000, 8000 and 16000.					

Tab. 3: *Subdivision of main types into types according to alkalinity, on a 2log-basis (adapted after STUYFZAND 1989c).*

Class	code	conditions for [Na+K+Mg]corr (meq/L)
[Na+K+Mg]deficit[1]	—	$< -\sqrt{1/2Cl}$ and $< 1.5(\Sigma K - \Sigma A)$
[Na+K+Mg]-equilibrium[2]	.	$> -\sqrt{1/2Cl}$ and $< +\sqrt{1/2Cl}$ and: $\|[Na+K+Mg]corr + \frac{(\Sigma K - \Sigma A)}{\|\Sigma K - \Sigma A\|} \cdot \sqrt{1/2Cl}\| > 1.5 \| \Sigma K - \Sigma A \|$
[Na+K+Mg]surplus[3]	+	$> +\sqrt{1/2Cl}$ and $> 1.5(\Sigma K - \Sigma A)$
[1] = often indicative of a salt water intrusion (anywhere, any time) [2] = often indicative of adequate flushing with water of constant composition [3] = often indicative of a fresh water intrusion (anywhere, any time)		

Tab. 4: *Subdivision of subtypes into 3 classes, based on [Na+K+Mg]corrected for sea salt (after STUYFZAND 1989c). $\Sigma K, \Sigma A$ = sum of cations and anions resp.*

type, subtype and class of a water sample, each of which contributes to the total code (and name) of the water type (fig. 1).

Chlorinity determines the main type, as indicated in tab. 2. Each main type is subdivided into 11 types according to alkalinity (tab. 3). For most natural waters with 4.5 < pH < 9.5 alkalinity equals $HCO_3 + CO_3$ in meq/L. The most important cation and anion determine the subtype in a specific way as outlined in STUYFZAND (1989c).

Finally, each subtype is subdivided into 3 classes (tab. 4) according to a parameter introduced by STUYFZAND (1986): the sum of Na, K and Mg in meq/L, corrected for a contribution of sea salt:

[Na+K+Mg]corr = [Na+K+Mg]measured - 1.0716 Cl

Significantly positive and negative values can often be associated with cation exchange due to fresh or salt water intrusion respectively. Values not deviating significantly from zero are indicative for equilibrium between the sea salts in solution and those on the exchanger, due to an adequate flushing of the porous medium with water of constant composition. Positive values may also be linked to dissolution of minerals or mineralization of biomass. Negative values may also point at the synthesis of minerals or biomass. For more details reference is made STUYFZAND (1988b and 1989c).

2.4 The redox index

The direct measurement of the redox potential (Eh) with electrodes runs into practical problems and is handicapped by unreliable results (LINDBERG & RUNNELLS 1984). Unfortunately the same holds for its calculation from a single redox pair like Fe^{2+}/Fe^{3+} (LINDBERG & RUNNELLS 1984). Therefore the semi-empirical redox indexing according to STUYFZAND (1988a) as outlined in fig. 2, is choosen here.

2.5 The pollution index

The Pollution Index proposed by STUYFZAND (1988a) is choosen here, because of its applicability to each hydrological compartment, ground water included. This index is based on six, equally weighted quality aspects: pH, [$NO_3 + SO_4$-corrected for a sea salt contribution], the sum of several trace elements in unfiltrated samples, including a weighting factor, chlorinated hydrocarbons adsorbable to activated coal (= AOCl), the amount of thermotolerant Coli bacteria (often zero in ground water), and tritium activity.

The nomenclature and examples of discerned pollution classes are given in tab. 5. More details are given in STUYFZAND (1990).

2.6 Identifying the origin and delineating hydrosomes

Tab. 6 lists the hydrosomes, which may be encountered in coastal plains. The identification of the origin of a ground water sample may pose a problem. Natural tracers obtained prior to infiltration (like Cl, Br, ^{18}O) or upon passage of a geochemically active subsoil layer, like lagoonal clay (HCO_3, DOC, I), form indispensable tools in addition to hydrological observations like the ground water flow pattern (STUYFZAND 1990).

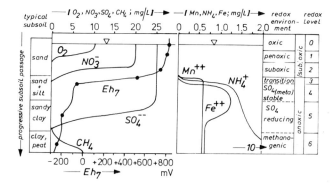

Fig. 2: *Classification of the redox environment, based on the presence or absence of main redox components of water (O_2, NO_3, SO_4, Fe, Mn and CH_4). Subsoil passage is assumed as plug flow in a system closed from the atmosphere and progressively richer in organic carbon. Original O_2, NO_3 and SO_4 concentrations are set at 10, 20 and 25 mg/L respectively. The indicative redox potentials at pH = 7 (Eh7) are derived from STUMM & MORGAN (1981). Slightly modified after STUYFZAND (1989b).*

name of pollution class	POLIN	example
unpolluted	<0.5	>100 years old, deep dune water
quasi unpolluted	0.5–1.5	shallow, calcareous dune water
slightly polluted	1.5–2.5	shallow, acid dune water
moderately polluted	2.5–3.5	bulk precipitation
polluted	3.5–4.5	pretreated Rhine water for AR
heavily polluted	4.5–5.5	river Rhine during low flow
extremely polluted	5.5–6.5	municipal sewage effluent
dead	>6.5	industrial sewage effluent
AR = Artificial Recharge		

Tab. 5: *Nomenclature of discerned pollution classes, with examples from The Netherlands. The pollution index POLIN has a logarithmic character.*

3 Application to the coastal area of Bergen, NW-Netherlands

3.1 Setting of the area

The coastal area of Bergen, circa 40 km northwest of Amsterdam, consists of coastal dunes, shallow polders and deep polders (reclaimed lakes; fig. 3). Major impacts of man started about 1000 AD, with the construction of dykes to prevent the invasion of the sea in the tidal flats and marshlands behind the dunes. Most shallow brackish lakes in the area were reclaimed since 1550 AD.

Since 1885 ground water is abstracted from the dune area for drinking water supply. The ground waters considered, are contained in (silty) sands, (sandy) clays and peat of Quarternary age, to a depth of circa 260 m below Mean Sea

Hydrochemical Analysis of Coastal Dunes and Low Lands

hydrosome			main
code	name	Origin	recharge mechanism
A	Artificial recharge	remote meteoric	pumps
B	river Banks	local meteoric	precipitation
C	Creek ridge	local meteoric	precipitation
D	coastal Dunes	local meteoric	precipitation
E	Estuary	remote meteoric + cycl. mar.	floods, pumps
F	Fluvial	remote meteoric	floods, pumps
J	Juvenile	juvenile	heat
L	Lagoon & tidal flats	cyclic marine + rem. meteor.	density, pumps
M	connate Marine	connate marine	absent, compaction
P	marsh land & Polder	local & remote meteoric	pumps, precipitation
S	open Sea, near shore	cyclic marine	density, pumps
U	remote Uplands	remote meteoric	precipitation
V	Vulcanic, mag & plut.	meteoric and/or connate	precipitation, heat

Tab. 6: *Major hydrosomes in coastal plains, in alphabetic order (after STUYFZAND 1990).*

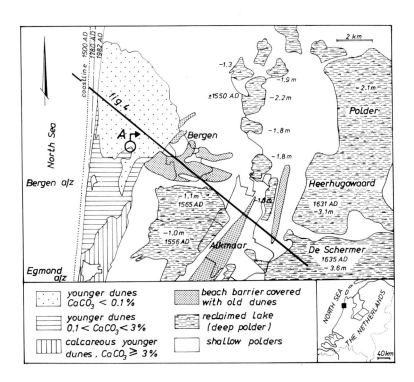

Fig. 3: *Location and physiographic subdivision of the coastal area of Bergen, with some historical developments during the past 500 years. The ground water abstraction area for drinking water supply is coded "A".*

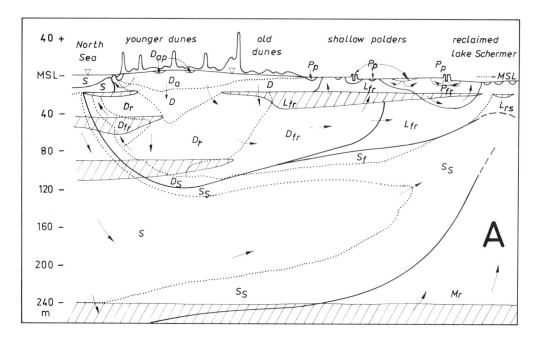
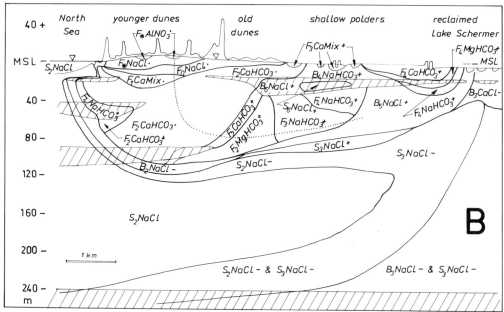

Fig. 4: Cross section of the coastal area near Bergen, NW-Netherlands, showing the areal distribution of hydrosomes with their hydrochemical facies (A) and of chemical watertypes (B). Simplified after STUYFZAND (1989e). For location of the profile see fig. 3, the hatched areas indicate the aquitards.

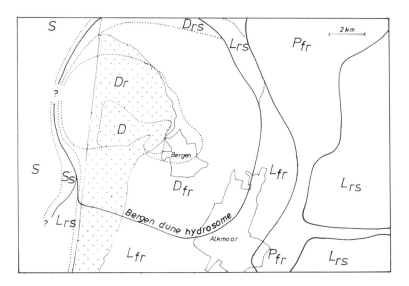

Fig. 5: *Horizontal section at a depth of 30 m-MSL over the coastal area near Bergen, NW-Netherlands, showing the areal distribution of hydrosomes and their hydrochemical facies in the (semi)confined, second aquifer (adapted after STUYFZAND 1989e).*

Level (MSL). For further details on the geology and hydrogeology reference is made to STUYFZAND (1988b, 1989e). Mean annual precipitation amounts to 750 mm. Mean temperature is 10°C.

3.2 Genesis and areal distribution of hydrosomes and their facies

The areal distribution of hydrosomes and their facies, without chemical watertypes for ease of survey, is presented in a cross section (fig. 4A) and in a horizontal section at about 30 m.-MSL (fig. 5). Watertypes are separately shown for the same cross section in fig. 4B, such that the main types fresh and fresh-brackish as well as brackish and brackish-salt have been associated into respectively fresh (F = 30 < Cl < 300 mg/L) and brackish (B = 300 < Cl < 10000 mg/L). For most discerned watertypes a typical chemical analysis is listed in STUYFZAND (1989d). Lack of variation justifies the omission of temperature. Hydrosomes have been coded in accordance with tab. 6, whereas boundaries between two hydrosomes are set at their 50% mixing divide.

The following hydrosomes have been recognized, in order of decreasing age:

The connate, marine Maassluis hydrosome (M)

The oldest hydrosome in the upper 260 meters of the studied area is formed by connate North Sea water in the Maassluis Formation, which was deposited at the beginning of the Pleistocene epoch some 2.10^6 years ago. Heavy drainage of the reclaimed deep lakes, causes this water to ascend there above its normal upper level at around 240 m.-MSL.

A unique combination of strong depletion of SO_4 and K with a relatively low alkalinity and small $\{Na+K+Mg\}$deficit respectively, distinguishes this hydrosome from the other brackish and saline ground waters. Chlorinity varies in between 8,900 and 13,500 mg/l. The facies is reduced and salinized leading to the code M_{rs}, with the watertypes B_3- to S_3-NaCl-.

The relict lagoon and tidal flat hydrosome (L)

These brackish and saline ground waters were formed 8000 till 300 years ago during the Holocene transgression. Current recharge is lacking due to protection of the area from the sea by manmade dykes and sluices.

Major extensions are situated east of the coastal dune hydrosome (fig. 5), and are bleeding out in seepage areas like Heerhugowaard and Schermer.

Most relict lagoon and tidal flat waters in the studied area can be recognized by their very high to extreme alkalinity (16–64 meq/L).

The facies is calcareous, reduced and, depending on its hydrological position, freshened or NaKMg-equilibrated or salinized, which results in the following codes: L_{fr}, L_r and L_{rs} respectively. The freshened hydrosome contains the watertypes $B_4 - B_6$NaHCO$_3$+, $B_4 - B_6$NaCl+ and $S_5 - S_6$NaCl+, the NaKMg-equilibrated hydrosome mainly $B_5 - B_6$NaCl., and the salinized hydrosome in the deep polders $B_3 - B_4$CaCl-, $B_4 - B_5$NaCl- and S_5NaCl-.

The North Sea hydrosome (S)

These saline ground waters have been formed since the Holocene transgression around 7500 years ago. During the first 7100 years the intrusion was driven by density mainly. Drainage of the land and marshes behind the dunes accelerated the intrusion considerably.

Intruded North Sea water can be easily recognized by its typical chlorinity in between 16,000 and 17,000 mg/l. The facies is calcareous, oxidized and NaKMg-equilibrated (S) with the watertype S_2NaCl., or calcareous, oxidized and salinized (S_s) with watertypes S_2– to S_3-NaCl-.

The coastal dune hydrosome (D)

The first coastal dune hydrosomes were formed around 5000 years B.P. (fig. 3). The small hydrosomes close to the reclaimed lakes Heerhugowaard and Schermer and below the cities Alkmaar and Heiloo vanished due to export of rain water by sewers in urbanized areas and due to displacement by polder water, which is conducted by canals and ditches through the old dune areas for irrigation of agricultural plots.

Around 1000 A.D. the formation started of the younger dunes, which were blown over the older dune ridges. Consequently, the fresh water lens eroded west of the present shoreline and expanded inland.

Later deformations of the fresh dune water lens are related to the very strong artificial drainage of the reclaimed deep lakes Heerhugowaard and Schermer (fig. 3), excessive ground water exploitation, a lowering of the water levels in the shallow polders, afforestation of the dunes and a continued coastal erosion of the northern dunes (fig. 3).

Coastal dune waters without admixing, can be recognized from fresh polder waters by their lower Cl content (30–60 as compared to 60–300 mg/L) and lower concentrations of SO$_4$ and/or HCO$_3$.

The following facies are subsequently encountered along a flow line from the surface in the northern dunes towards the intrusion front against relict tidal flat water at 70 m-MSL in the shallow polders (fig. 4A):

- acid, polluted, NaKMg-equilibrated and oxidized $F_*AlNO_3.$, $(=D_{ap})$, in the upper two meters of ground water;
- acid, NaKMg-equilibrated and oxidized F_*- and $F_oNaCl.$ mainly, $(=D_a)$, occurring below the acid, polluted facies down to a maximum depth of 20 m-MSL;
- calcareous, NaKMg-equilibrated and oxidized $F_1CaMix.$ and $F_2CaHCO_3.$, $(=D)$, in the main recharge area of the (semi)confined, second aquifer in the centre of the Bergen dune hydrosome, below 20 m-MSL to a maximum depth of 50 m-MSL (fig. 4–5).
- calcareous, NaKMg-equilibrated and reduced $F_2CaCO_3.$, $(=D_r)$, which surrounds the former oxidized facies (D) in the upper aquitard and (semi)confined second aquifer (fig. 5) to a maximum depth of 100 m-MSL in the Bergen dune hydrosome (fig. 4).
- calcareous, freshened and reduced $(=D_{fr})$, which forms the outer shell of the Bergen dune hydrosome, on its eastern sides and on the southern and northern sides (fig. 4 and 5). The typical zonation of the watertypes F_2CaHCO_3+, F_2MgCO_3+, F_3- to F_4NaHCO_3+ and B_3- to B_4NaCl+ within this group of facies, is discussed by STUYFZAND (1986, 1988b).

And finally there is a calcareous, reduced and salinized facies $(=D_{rs})$, present as the outer shell of the Bergen dune hydrosome, at least on its western and northern sides (fig. 5) and at its base (fig. 4). The typical zonation of the watertypes F_2CaCl-, B_2CaCl- and B_2- to B_3NaCl- within this group of facies, is discussed by STUYFZAND (1990).

The polder hydrosome (P)

In the course of the last 1000 years man created an intricate, dense drainage system in polders behind the dunes, consisting of canals, ditches and drains. The main canals have the highest water levels often well above the polder surface. They are also flushed and rinsed with water partly stemming from the river Rhine, in order to reduce the salinity of the surface water, which is supplied to the polders in the dry season.

Water from such canals and both surface and rain water in shallow polders adjacent to deep polders, may recharge underlying aquifers and form a rather extensive polder hydrosome, like to the west of the deep reclamations Heerhugowaard and Schermer (fig. 4 and 5).

As a consequence of variable contributions of brackish seepage water, Rhine water and local rain water as well as a highly variable land-use, no polder water is alike. However, the large hydrosomes are surely dominated by infiltrating canal water, whose composition is well known, and can be identifed on their Cl, SO_4 and HCO_3 content.

The facies of the discerned hydrosomes is either calcareous, freshened, oxidized and polluted $F_3CaMix+$, $(=P_{fp})$, or calcareous, freshened and reduced F_4CaHCO_3+ or F_4MgHCO_3+ or F_4NaHCO_3 $(= P_{fr})$, depending on the stage of freshening.

4 Concluding remarks

With the hydrochemical facies analysis proposed, the extremely complex hydrochemical situation in the coastal area of Bergen in the NW-Netherlands has been unravelled to a satisfactory degree. The results may satisfy hydrologists looking for the main recharge areas, for a chemical visualisation of flow patterns and for the present stage of evolution of flow systems; hydrochemists interested in modelling and searching for the best flow lines and all relevant processes that must be taken into account; and ecohydrologists studying natural or anthropogenic changes in flora and fauna, which may be linked to changes in saturation, redox or pollution index, the chemical water-type or origin.

Certainly, many data were required and this may not always be feasible. The whole procedure should be flexible enough to cope with lacking data then, for instance by cutting out facies-parameters for which data are missing, and simplify. On the other hand, it may be desirable to further differentiate in monotonous situations or in ecohydrological studies, for instance by assigning each redox or pollution class, or different classes of mineral saturation indices as a facies-parameter.

References

BACK, W. (1960): Origin of hydrochemical facies of ground water in the Atlantic Coastal plain. Internat. Geol. Cong. 21st, Copenhagen 1960, Rept.pt.1, 87–95.

LINDBERG, R.D. & RUNNELLS, D.D. (1984): ground water redox reactions: an analysis of equilibrium state applied to Eh measurements and geochemical modelling. Science **225**, 925–927.

PLUMMER, L.N., JONES, B.F. & TRUESDELL, A.H. (1976): WATEQF, a Fortran IV version of Wateq, a computer program for calculating chemical equilibrium of natural waters. U.S. Geol. Surv. Water Res. Invest. **76-13**, 61 pp.

STUMM, W. & MORGAN, J.J. (1981): Aquatic chemistry, an introduction emphasizing chemical equilibria in natural waters. J. Wiley & Sons, NY, 2nd. ed., 780 p.

STUYFZAND, P.J. (1984): Groundwater quality evolution in the upper aquifer of the coastal dune area of the western Netherlands. IAHS Publ. **150**, 87–98.

STUYFZAND, P.J. (1986): A new hydrochemical classification of water types: principles and application to the coastal dunes aquifer system of the Netherlands. Proc. 9th Salt Water Intrusion Meeting, Delft 12–16 may, Delft Univ. Techn., 641–655.

STUYFZAND, P.J. (1988a): Alkalinity, a redox and pollution index as parameters and options in a hydrochemical classification of water types. H_2O (21)22, 640–643, in dutch.

STUYFZAND, P.J. (1988b): Hydrochemical evidence of fresh and salt water intrusions in the coastal dunes aquifer system of the Western Netherlands. (Flemish) Natuurwetensch. Tijdsch. **70**, 9–29.

STUYFZAND, P.J. (1989a): An accurate, relatively simple calculation of the saturation index of calcite for fresh to salt water. J. Hydrol. **105**, 95–107.

STUYFZAND, P.J. (1989b): Hydrology and water quality aspects of Rhine bank ground water in The Netherlands. J. Hydrol. **106**, 341–363.

STUYFZAND, P.J. (1989c): A new hydrochemical classification of water types, with examples of application. IAHS Publ. **182**, 89–98.

STUYFZAND, P.J. (1989d): Factors controlling trace element levels in ground water in The Netherlands. Proc. 6th Water Rock Interaction Symp., Malvern (UK), 3-8 aug. 1989, D.L. Miles (ed.), A.A. Balkema, Rotterdam, 655–659.

STUYFZAND, P.J. (1989e): Hydrochemistry and hydrology of dunes and adjacent polders in between Egmond and Petten, NW-Netherlands. KIWA-report SWE. 87-001, in dutch, 239 p.

STUYFZAND, P.J. (1990): A method of hydrochemical facies analysis: I. theory. Submitted to J. Hydrol.

ZAGWIJN, W.H. (1986): The Netherlands in the Holocene. In: Geology of The Netherlands (I), RGD, Staatuitgeverij, The Hague, 46 p. (in dutch).

Address of author:
P.J. Stuyfzand
The Netherlands'Waterworks'Testing and Research Institute KIWA Ltd.
P.O. Box 1072
3430 BB Nieuwegein
The Netherlands

GEOHYDROLOGY OF LES DUNES DE MONT SAINT FRIEUX, BOULONNAIS, FRANCE

T.W.M. **Bakker** & P.R. **Nienhuis**, Alkmaar

Abstract

A description is given of the geohydrology of Les Dunes de Mont Saint Frieux. The dunes in the study area overlie a transition zone between a high limestone plateau in the east and a Pleistocene marine abrasion platform to the west. Most of the groundwater in the dune-area is derived from subsurface recharge from the limestone area, only a minor part comes from direct infiltration by precipitation. Consequently, the groundwater composition is dominated by calciumcarbonate. Because of the dependence on recharge from the high limestone areas, the water quality in the dunes is vulnerable for pollution caused by human activities on the plateau.

1 Introduction

Along the northeastern coast of France, a range of extensive dune areas can be found between the mouth of the river Somme and the French / Belgian frontier. The geohydrology of Les Dunes de Mont Saint Frieux, a nature reserve covering some 5 square km. (fig. 1), has been studied in the course of an ecological inventory for a management master plan (VAN GENDEREN E.A. 1989). This dune area portrays some striking features not often found in coastal dune areas. First, it forms the transition zone between a marine abrasion platform and a limestone plateau, the latter reaching an altitude of over 150 m above m.s.l. Second, its war history precludes the installation of groundwater monitoring wells commonly used in geohydrological studies. These factors, together with the limited time available, led to the use of methods derived from a technique called hydrological systems analysis. This technique is based on a theory of TOTH (1963) and has further been developed in the Netherlands by Engelen (see ENGELEN 1985, ENGELEN & JONES 1987). It makes use of information on virtually every aspect of the landscape which can be related to hydrology. Data on geology, geomorphology, soil cover, groundwater quality and quantity, climatology, vegetation, land use and even history and (historical) topographic names are combined in order to arrive at a coherent picture of the geohydrology.

In the following we give a survey of these aspects in Les Dunes de Mont Saint Frieux.

ISSN 0722-0723
ISBN 3-923381-23-9
©1990 by CATENA VERLAG,
D-3302 Cremlingen-Destedt, W. Germany
3-923381-23-9/90/5011851/US$ 2.00 + 0.25

Vegetation type	Evapotranspiration (mm/year)	Net Precipitation (mm/year)
Bare dunes	180	540
Wet dune valleys	550	170
Dry dune vegetation	360	360
Humid deciduous forest	550	170
Dry deciduous forest	400	320
Humid pine forest	700	20
Dry pine forest	550	170

Tab. 1: *Evapotranspiration values from some vegetation types under average climatic circumstances in Les Dunes de Mont Saint Frieux. (Derived from BAKKER 1981).*

2 Climate

The local climate strongly resembles that of other dune areas elsewhere in Western Europe. Westerly winds prevail and meteorological lows bring in most of the annual recharge. We deduced the average annual precipitation, 720 mm, from two meteorological stations nearby to the north and south, of 620 and 817 mm/y, resp. The evapotranspiration is computed using a method proposed by BAKKER (1981) for the Netherlands. In tab. 1 the evapotranspiration for some representative vegetation types, which can be found in Les Dunes de Mont Saint Frieux, has been summarized.

3 Geology

The Boulonnaise area of which Les Dunes de Mont Saint Frieux are a part, is underlain by the Artois anticline. The axis of this anticline runs roughly east-west from the Ardennes to southwestern England (fig. 1). The upper part of it has been strongly eroded. Nowadays the surface of this part is lower than its surrounding, and is called Basse (= Lower) Boulonnais. Escarpments of a relatively height up to 150 m above m.s.l., mark the border with the Haute (High) Boulonnais. This plateau is formed by Cretaceous marls and marly limestones. The study area is situated on the southern limb of the anticlinal, just south of the border between the Basse and Haute Boulonnais. The stratigraphy and the geological build-up of the area are depicted schematically in fig. 2. Cretaceous limey deposits are exposed in the eastern and northeastern part and form high cliffs, of which the Mont Saint Frieux is the highest (153 m above m.s.l.). A somewhat lower ridge is situated along the eastern border. On aerial photographs it can be seen that the location of the cliffs is governed by zones of structural weakness (e.g. faults and joints; see fig. 3).

To the west the Cretaceous is overlain by Holocene dune deposits. The thickness of this sand cover is largely unknown. The same applies for what follows directly underneath. No outcrops of limestone have been found in the proper dune area, and of the only known (public drinking water supply) well in the vicinity no well description is available. BRIQUET (1930) distinguishes older and younger dunes, the latter dating from after the beginning of the present era. In the subsurface some peat layers are present. Relatively continuous lower peat layers are possibly related

Geohydrology of Les Dunes de Mont St. Frieux, France

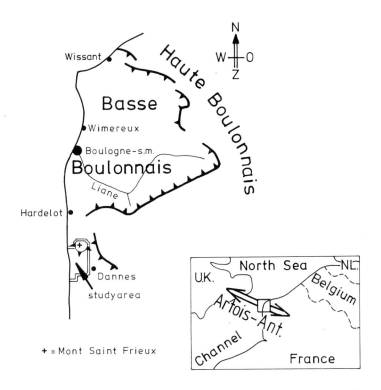

Fig. 1: *Location of Les Dunes de Mont Saint Frieux on the southern limb of the Artois-anticlinal.*

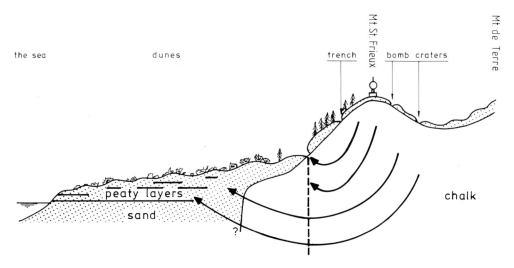

Fig. 2: *Simplified geological cross-section of the area with a scheme of the main groundwater flow.*

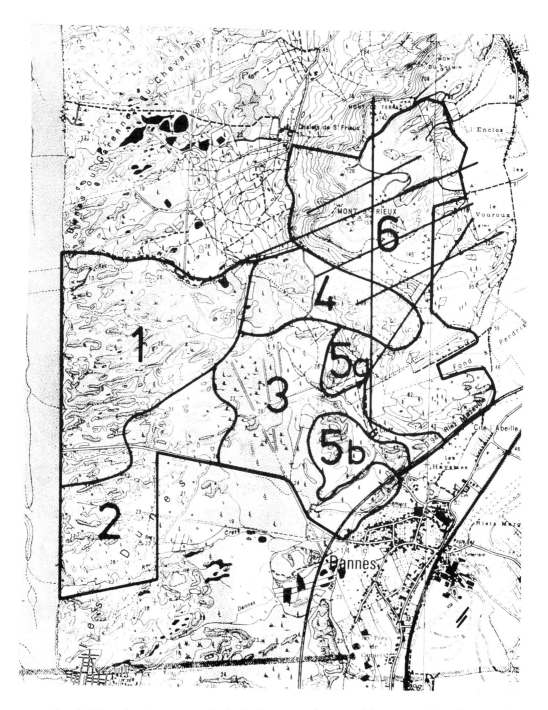

Fig. 3: *Map showing geomorphological zones and trace of lineaments based on aerial photo interpretation.*

to eustatic sea level movements during the Holocene. Less continuous shallower peat layers possibly have developed in dune valleys during cycles of warping and burial by drift sand. Bog peat exposures along the beach have been dated at 4000 B.C. (BRM 1985). The transition zone between the old marine abrasion platform and the high cliffs is marked by steep sloping sand dunes. A thin veneer of sand has also been found on top of some of the cliffs.

BRIQUET (1930) suggested that a now buried Pleistocene coastline runs in a NNW-SSE direction through the study area, to the west of the Mont Saint Frieux. This ancient shoreline might be interpreted as the border between shallow limey deposits in the east and a possibly thick sand cover in the west. However, given the thick sand cover and the clearly structurally controlled cliff morphology, the exact trace of this shoreline is, according to BRIQUET, a matter open to speculation.

4 Composition of the dune sand

Dune sand samples from the area merely consist of quartz grains and contain very little material from the Cretaceous. The primary lime content is only a few percent. The limey part merely comprises a dolomitic weathering residue, which is more resistant to weathering than pure limestone derived from shell grit (aragonite). Dune soils containing shell grit are buffered against acidification as long as the lime content is more than 0.3 percent (KLIJN 1981). In the case of Les Dunes de Mont Saint Frieux, acidification may occur already at higher lime contents, due to the more inert characteristics of these lime fragments. From this perspective it is not surprising that the vegetation in the older dunes suffers slight acidification.

5 Geomorphology

Field surveys and aerial photo interpretation of the dune area proper revealed the following:

- secondary dune shapes, merely comprising single parabolic dunes or blown out structures, predominate;

- the level of valley-bottoms increases strongly in landinward direction. This is caused by the position of the underlying deposits and the steep gradient of the groundwater-table;

- the surface water has a large influence on the geomorphology. Stream systems induced remarkably straight dune ridges in a zone just behind the fore dune.

Based on additional surveys the study area can be divided into the following geomorphological zones (fig. 3):

- zone 1, which until some years ago has been strongly influenced by the wind. The sand contains lime. Many wet valleys and dune streams are associated with large amounts of upward seeping fresh water. Parabolic dunes can be found along with some long, straight dune ridges; the latter are associated with small streams;

- zone 2, which also has been influenced by wind drift until very recently, some parts even until the present. Parabolic dunes prevail. The lime content is about the same as that of zone 1. However, the groundwater has had less influence here, and consequently less wet valleys and streams are present.

- zone 3, gently sloping westward, which has been overgrown for some time and contains large areas occupied by forest. Pedogenesis has occurred. The lime content varies strongly from place to place. Upward seeping fresh water is responsible for wet valleys and streams, too;

- zone 4 comprises merely dry, sparsely vegetated dunes which have recently been deposited on the westward sloping limey substratum (climbing dunes). The thickness of the sandcover is such that the limestone is completely covered. The sand has been redeposited and has a lower lime content than in the other zones;

- zone 5, a high dune area with a thick sand cover. This area is quite dry and forested with pine-trees. The lime content of the sand varies from place to place;

- zone 6. Here the limey substratum is locally exposed because the sand cover is on the whole relatively thin. The lime content again varies strongly. The area is dry and is covered with grass and shrubs. Some pine-tree forest has been planted. Dune deposits can be found up to 150 m above m.s.l. or more.

6 Hydrology

The limestone plateau east of the study area, the Mont Saint Frieux, the other limestone hills and the high eastern dune zone are all quite dry. This is related to their altitude and the high permeability of both the sand and the bedrock. This infiltration area is drained by various seepage zones and some springs directly at the foot of the hills. Some springs are permanent and feed streams which have a great influence on the dune area proper. The groundwater flow in a cross-section perpendicular to the coast line is schematically depicted in fig. 2. The dune area itself is recharged by means of three mechanisms: subsurface flow from the limey infiltration areas, infiltration of water in streams which are fed by springs and seepage zones and direct infiltration of precipitation. In order to make a contour map of the water table (fig. 4) we made use of the fact that the dune vallies cannot be blown out deeper than the groundwater level, and that the water table is very smooth and continuous due to the high permeability of the dune sand. The surfae water in ponds and small lakes can therefore be viewed as outcrops of the groundwater, and their water heights as gauges of the water table elevation. Because the altitudes of these water levels were deduced from topographic maps, the elevations are expected to show an error of at most 2 meters; the average error will be much less because of leveling out effects.

The groundwater flow pattern, deduced from the distribution of high and low areas and the groundwater table map, is shown in fig. 5.

7 Modeling

The water table elevations were input in a two-dimensional groundwater flow model which has been developed at the Free University in Amsterdam, Netherlands (VAN ELBURG E.A. 1987). Given the head distribution(s) at

Fig. 4: *Contour map of the groundwater table (contours in meters above m.s.l.).*

Fig. 5: *Sketch map of spring zones and the direction of groundwater flow.*

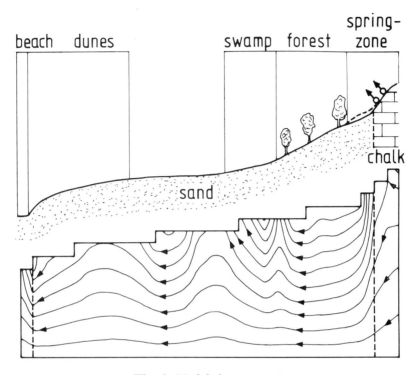

Fig. 6: *Modeled cross-section.*

open (flux) boundaries and hydraulic parameters, this model computes the internal head distribution and flowlines and -times in a cross-section. The (horizontal) permeability of the dune sand, averaged over its thickness, was assumed to be 10 m/day. It was also assumed that below the peat layer at mean sea level there exists another sand layer with a thickness of 15 m. Thus the thickness of the sandy aquifer decreases in eastward direction from 60 to 20 m. The permeability of the limestone was taken to be 15 m/day horizontally and 1 m/day vertically; the model proved to be relatively insensitive to the latter values.

With the model (fig. 6) we calculated a recharge from the limestone areas to the dune area in the order of 20,000 m^3/day. The dune area totals about 600 ha. If the average pecipitation surplus equals 1 mm/day, the recharge from precipitation amounts to 6000 m^3/day. This clearly shows that the subsurface recharge from the high limey infiltration areas to the east is more important than the recharge from direct precipitation.

The wedge-shape of the dunesand aquifer is clearly responsible for the wet conditions just behind the foredunes: to the west, the thickness of the aquifer becomes too small to accomodate for the total groundwater flux input by recharge from the limestone areas and precipitation. Consequently, the groundwater surplus is drained from there by various small dune streams which originate in this region.

8 Hydrochemistry

65 water samples were taken and analyzed in the field with simple field equipment; five more samples were analyzed in a laboratory. The results indicate that:

- the chloride contents (30 to 110 mg/l) show the high values usually found in coastal dune areas. BAKKER (1981) found values ranging from 30 to 100 mg/l, with outliers of 200 mg/l, in the Dutch coastal dunes;

- the concentrations of the dominant ions calcium and bicarbonate range from 75 to 140 and from 200 to 450 mg/l, resp., and clearly show the origin of most of the ground water in the dunes;

- some samples have high concentrations of nitrate (up to 40 mg/l). Concentrations from 5 to 10 mg/l are quite usual in dune areas. Higher contents indicate influence from antropogenic factors.

8.1 Spring water

All springs situated at the foot of the high limestone hills yield calciumcarbonate water, usually oversaturated with respect to calcite. The electrical conductivity is about 500 μS/cm. The nitrate contents are strikingly high (20 to 40 mg/l). This and the relatively high chloride contents point to pollution from fertilizers on the high limstone plateaus.

8.2 Stream water

The water in the streams originates from the forementioned spring zones and, consequently, its chemical composition in the upstream reaches mimics that of the spring water. As soon as the water enters a more or less stagnant reservoir (pond or small lake), the nitrate contents decrease strongly, presumably because of transfer into (free) N_2 by water plants (SCHLEGEL 1974, BLACKBURN 1983, DUEL & SARIS 1986). The EC-values are the highest in the spring zones and close to the coast.

8.3 Shallow groundwater

All samples are calcium-bicarbonate waters. In only a few samples we found clues for influence of precipitation. The shallow groundwater in the wet valleys close to the foredunes has EC-values of 500–600 μS/cm. The lower EC's between the spring zones and the foredunes are presumably due to dilution of the groundwater by precipitation in the dune area. Closer to the coast the effects of sea spray become evident. The surface water in the pine-forested areas yields higher EC-values (600–800 μS/cm). This is probably caused by interception of aerial matters by pine trees and by their relatively high evapotranspiration rate, the latter leading to higher concentrations in the remaining shallow groundwater.

9 Hydrobiology

SMIT (in VAN GENDEREN E.A. 1989) investigated the water mites in the study area. He found 58 species and concluded that Les Dunes de Mont Saint Frieux have a rich population of water mites when compared with two Dutch dune areas. Species which are typical for the Dutch dune areas are missing altogether; on the other hand, some species were found which are typical for running water in seepage areas. In addition, SMIT states that zoogeographic factors are less important than hydrological ones for an explanation of this dissimilarity.

10 Relation with the limestone recharge area

The evidence discussed above illustrates the dominating influence of the recharge from the high limestone areas, either by subsurface flow or indirectly by streamwater. This implies that all kind of human activities (e.g., fertilizers, pollution) in these recharge areas may have an impact on the groundwater of the dune area. A proper management of the dune area of Mont Saint Frieux must take this factor into account.

References

BAKKER, T.W.M. (1981): Nederlandse Kustduinen: Geohydrologie. Thesis, Pudoc, Wageningen, 189 pp.

VAN GENDEREN, J., TEN HAAF, C., BAKKER, T.W.M. & NIENHUIS, P.R. (1989): Les dunes de Mont Saint Frieux, plan d'amenagement et de gestions. Ten Haaf & Bakker, Alkmaar.

BLACKBURN, T.H. (1983): The Microbial Nitrogen Cycle. In: W.E. Krumbein (ed.), Microbial Chemistry. Blackwell, Oxford, 63–89.

BRIQUET, A. (1930): Le Littoral du Nord de la France et son Evolution Morphologie (suivi d'un Appendice: L'evolution du Rivage du Nord de la France et l'activité de l'homme). Librairie Armand Colin, Paris, 440 pp.

BRM (1985): Carte Geologique de la France a 1/50.000, feuille Boulogne-sur-Mer, avec Notice, 26 pp. Bureau de Recherches Géologiques et Minières, Orléans.

DUEL, H. & SARIS, F.J.A. (1986): Waterzuivering door Macro-heliofytenfilters. Landschap, 1986, no. **4**, 295–305.

ELBURG, H. van, HEMKER, C.J. & ENGELEN, G.B. (1987): FLOWNET Users Manual. Int. Rep. Free University, Amsterdam.

ENGELEN, G.B. (1985): Hydrological Systems Analysis. A Regional Case Study. TNO-DGV Institute of Applied Geosciences, Delft, rept. OS 84-20.

ENGELEN, G.B. & JONES, G.P. (1987): Developments in the Analysis of Groundwater Flow Systems. IAHS publ. no. **163**.

KLIJN, J.A. (1981): Nederlandse Kustduinen. Geomorfologie. Thesis, Pudoc, Wageningen, 188 pp.

SCHLEGEL, H.G. (1974): Allgemeine Mikrobiologie. G. Thieme Verlag, Stuttgart, 461 pp.

TOTH, J. (1963): A Theoretical Analysis of Groundwater Flow in Small Drainage Basins. J. Geoph. Res. **68**, 4795–4812.

Address of authors:
T.W.M. Bakker
P.R. Nienhuis
Ten Haaf & Bakker
Consultency for Dune-Management
Nieuwpoortslaan 61
1815 LK Alkmaar
The Netherlands

NEW PUBLICATION

Frank Ahnert (Editor)

LANDFORMS AND LANDFORM EVOLUTION IN WEST GERMANY

Special Publication on the Occasion of the SECOND INTERNATIONAL GEOMORPHOLOGICAL CONFERENCE, Frankfurt a.M. September 3 - 9, 1989

CATENA SUPPLEMENT 15

hardcover/352 pages/numerous figures, photos and tables

ISSN 0936-2568/ISBN 3-923381-18-2

list price: DM 89.-/US $ 49.-/ subscription price and standing order price CATENA SUPPLEMENTS: DM 62,30 /US $ 34.30

ORDER FORM

☐ Please send me .... at the rate of DM 89.-/ US $ 49.- .... copies of CATENA SUPPLEMENT 15.

☐ I want to subscribe to CATENA SUPPLEMENTS starting with no. 1 (30% reduction on the list price)

Name ....

Address ....

Date ....

Signature: ....

Please charge my credit card: ☐ MasterCard/Eurocard/Access ☐ Visa ☐ Diners ☐ American Express

Card No.: .... Expiration date: ....

Please, send your orders to:

CATENA VERLAG, Brockenblick 8, D-3302 Cremlingen-Destedt, West Germany, tel.05306-1530, fax 05306-1560

USA/Canada:**CATENA VERLAG**, Attn. Denize Johnson, P.O.Box 368, Lawrence, KS 66044, USA, Tel. (913) 843-1234, fax (913) 843-1244

GEOHYDROLOGY OF THE EOLIAN SANDS OF THE DOÑANA NATIONAL PARK (SPAIN)

M.R. **Llamas**, Madrid

Summary

Diverse and interesting ecosystems are located in the eolian sands of Doñana. Their diversity is related to factors such as wind, general geomorphology, precipitation and evapotranspiration, vegetation, and depth and fluctuation of the water table. In this paper the emphasis is placed on the role of groundwater. From the hydrological point of view, the eolian sands only form the upper part, in its recharge area, of a single large aquifer system kown as the Almonte-Marismas Aquifer (extending over an area of about 3400 km^2). This aquifer system plays a fundamental role in the functioning of the diverse habitats of the lower Guadalquivir Valley. The results of several digital numerical flow models indicate an increasing and serious environmental impact to the National Park of Doñana (760 km^2) caused by the abstraction of groundwater for water supply and irrigation.

1 Introduction

The main aim of this paper is to present the role of hydrogeological conditions in the existence and evolution of the different habitats or ecosystems located in the sands at the National Park of Doñana, near the mouth of the Guadalquivir River, Spain. Knowledge of the hydrological functioning of such diverse ecosystems seems crucial if they are to be conserved through proper management.

The lack of practical knowledge about the hydrogeological factors in the ecology of wetlands may lead to serious damage to these ecosystems. This problem is probably most serious in arid and semiarid countries where the demand of fresh groundwater for irrigation is usually great. The groundwater system of the Doñana National Park forms part of the general hydrogeological system of the lower Guadalquivir Valley which covers an area of 3400 km^2.

2 General characteristics

The Doñana National Park (DNP) is probably the most important natural monument in Spain and one of the rincipal ones in Europe. It was created in 1969 and covers an area of 760 km^2 (fig. 1). Its current legal status was enacted by the Spanish Government in 1978. In 1979 the Spanish Government also approved the "Plan de Transformación Agraria de Almonte-Marismas" (PTAAM) which was supposed to become the most important irrigation project using groundwater in

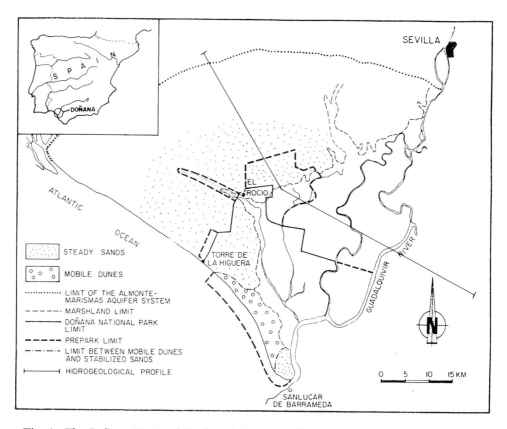

Fig. 1: *The Doñana National Park and the lower Guadalquivir aquifer modified from LLAMAS (1988).*

Spain: 24000 ha with a pumpage of 145 hm^3/year. The PTAAM area is located on the same geomorphological and hydrogeological system as the DNP.

3 Climate

The mean annual precipitation is about 600 mm. Its seasonal and interannual variation is great; 80% of the rainfall occurs between October and March. The average annual temperature is 18–19°C. Potential evapotranspiration is estimated at about 900 mm/year. Actual evapotanspiration depends largely on the precipitation, and consumes between 50% and 100% of the rainfall in humid and dry years, respectively. Free surface water evaporation is about 1500 mm/year.

4 Geology

The DNP and the PTAAM are located in the lower valley of the River Guadalquivir. This area, (fig. 1), forms a large plain that was originally an estuary and is now filled with Pliocene to Quaternary sediments with a rather heterogeneous lithology (fig. 2a). The top of most of these deposits — except the muds

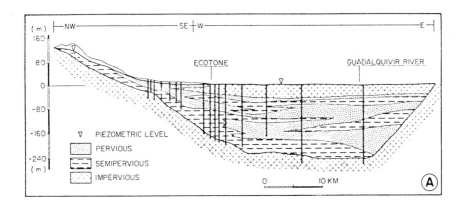

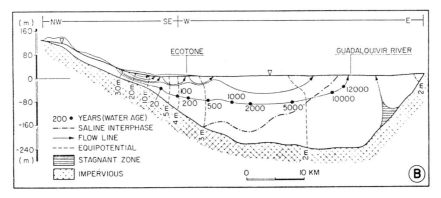

Fig. 2: *(a) General hydrogeological profile; (b) Simplified Groundwater Flow System, Location of the profile in fig. 1 (Modified from RODRIGUEZ & LLAMAS 1986).*

of the central part — is covered with Holocene eolian sands. The most recent sands form moving dunes, whereas the older sands are stabilized by vegetation (fig. 1). The thickness of the eolian sands is not greater than 20–30 m and is frequently no more than a few meters only. The central part of the area is covered by a layer of estuarine and marshy clays of low permeability with thicknesses up to 100 m. The Pliocene to Quaternary materials constitute an extensive detritic aquifer system (3400 km²). The aquifer is underlain by Miocene marine marls, considered to be impervious. The maximum depth of this impermeable base is over 200 lm below the surface (fig. 2).

5 Geomorphology

Three major landscapes in the doñana area have been identified:

1. the steady eolian sands;

2. the moving dunes; and

3. the marshes (fig. 1).

The first two are located on eolian deposits, the last on the marshy clays.

The Doñana's system of moving dunes is one of the most spectacular landscapes in Spain and in Europe. It extends from the mouth of the Guadalquivir river to Torre de la Higuera (fig. 1). The moving dunes cover an area 30 km long and three to five km wide. Four or five longitudinal dune ridges parallel to the coast and 20 to 30 m high are continuously migrating towards the NE, with an average speed of 4 to 5 m/year. Vegetation — mainly pine-trees — grows in the depressions ("corrales") between the longitudinal dunes. where the watertable is close to the surface. As the dunes advance they bury and destroy the vegetation. A new biocoenosis is set up after the passage of the dune. In the depressions or periphral to the moving dunes there may exist ephemeral or permanent lagoons.

The stabilized sands or "cotos" (fig. 1) are old dunes, now usually flattened by erosion, although sometimes a certain undulation can still be observed.

The marshlands cover the vast delta formed behind the littoral dunes (fig. 1). What is most impressive from the scenic point of view is their extreme flatness. However, on a microrelief scale, differences in altitude of a few centimeters occur.

6 Geohydrology of the DNP

In the eolian sands the vertical distance between the topographical surface and the watertable seems to be an significant factor for determining the type and density of the vegetation: the "Monte blanco" type where this depth to the watertable in September is greater than 1.5 m and the "Monte negro" type where the depth is smaller than 1.5 m (ALLIER et al. 1974). MERINO & MERINO (1988) consider that the vegetation of the "cotos" is strongly dependent on the depth of the watertable and that its depletion could damage the whole ecosystem. Small lagoons, seepage zones and small springs ("caños") exist in the downs ("corrales"), in places where the watertable rises above the topographical surface. These lagoons and "caños" are more frequent and permanent at the ecotone or contact line between two ecosystems, in this case between the peripheral eolian sands and the central estuarine clays. This ecotone at the inland side of the dunes is really a "wet-meadow".

During the rainy season almost the entire area of the marshlands is covered by a layer of water of an average depth of 30 cm. In summer virtually the whole area dries out and presents a typical steppe landscape with saline soils and halophytic and xerophytic vegetation. The microrelief of the marshlands controls the duration of the flooding, the quality of the water and the plant and animal biocoenosis.

The ecotone is the most fertile and productive zone of the National Park as a result of its permanent humidity and of the fertilization that it receives from the animals living there or permanently crossing it. This permanent humidity of the ecotone is caused by the slow but permanent natural upward discharge of groundwater as a consequence of its situation within the major Almonte-Marismas aquifer. In other words, the functioning of the ecotone can only be understood within the general hydrogeological framework of the large aquifer system of the lower Guadalquivir valley.

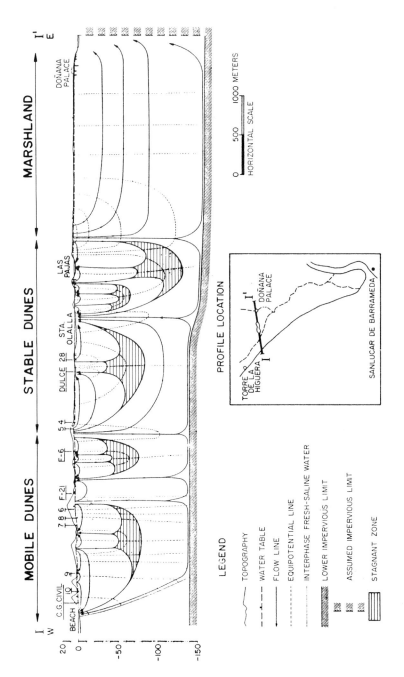

Fig. 3: *Numerical model of the dunes groundwater flow system (after VELA & LLAMAS 1986).*

7 Geohydrology of the Almonte-Marismas aquifer

Beneath the lower Guadalquivir valley there is an extensive detritic aquifer (3400 km^2). The peripheral zone of this aquifer (1900 km^2) is unconfined, a great part of it being covered with eolian sands, and receiving direct recharge from rainfall. The central parts of the aquifer remain confined beneath the muds of themarshlands (fig. 1 and 2a).

The aquifer transmissivity ranges from 10^{-4} m^2/s ($\approx$ 10 m^2/d) in the peripheral area (thin eolian sands) up to more than 2×10^{-2} m^2/s ($\approx$ m^2/d) in the central area (gravels and sands 150 m thick). A numerical steady state model (fig. 2) indicated the probable existence of several local flow systems within the unconfined part of the aquifer.

Fig. 3 shows the results of similar numerical simulations by VELA & LLAMAS (1986). This model simulates the flow system in a short representative profile (5 km) corresponding to the moving dunes area. It also assumes a steady state flow. Obviously, such assumption does not correspond to reality because of the changing boundaries of the topographic surface (the dunes advance about 4 m/year) and of the seasonality in the recharge. Also the distribution of the different geological materials (eolin sands, marine sands, estuarine muds, gravel, etc.) is not well known and therefore has been simplified in the model. Although, consequently, the flow net is only an initial approximation, its analysis shows several interesting features. It suggests the existence of both local and regional flow systems. The residence time of groundwater in the aquifer ranges from a few months to more than 200 years. The groundwater discharge occurs mainly at the depression to the permanent or ephemeral lakes or to the vegetation at the "corrales".

The chemical characteristics of the groundwater in the eolian sands are generally uniform. It is calcium bicarbonate water with a salinity usually lower than 500 ppm. Nevertheless, at the discharge areas (lakes of "corrales") the chemical composition of surface and groundwater may be quite different because of the complex and transient (seasonal) relationship between surface and groundwater and of the influence of biochemical processes.

8 Impact of exploitation of groundwater on the eolian sands

The previous part of this paper shows a clear relation between the different ecosystems located in the eolian sands and the occurrence and movement of groundwater. Of course, there still exists a good number of uncertainties about the detailed hydrological processes involved, mainly related to the unsaturated-saturated limit and its fluctuation. Nevertheless, in this author's opinion, the existing data, at least since 1982, are sufficient to predict a clear conflict between the groundwater development plans promoted by the Spanish Government and the conservation of the Doñana National Park which is also a clear responsibility of the Spanish Government. This conflict has not been settled yet; but this article does not seem to be the appropriate place for a discussion of this subject. The interested reader is reffered to RODRIGUEZ & LLAMAS (1986), HOLLIS et al. (1989), or LLAMAS (1988a, b, 1989a, b).

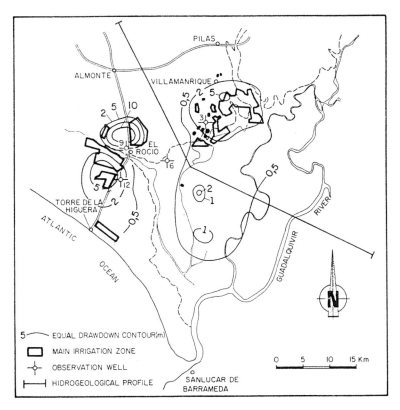

Fig. 4: *Computed watertable drawdowns after 40 years of constant pumping (45 hm^3/year) for irrigation of about 5600 ha, after IGME (1982) numerical model (modified from RODRIGUEZ & LLAMAS 1986).*

RODRIGUEZ & LLAMAS (1986) warned that, according to the basic results of a numerical flow model run by the Spanish Geological Institute (IGME) in 1982 (fig. 4) and according to the observed watertable depletion, the irrigation of only 5600 ha would produce a serious ecological impact in the Doñana National Part. At that time the total irrigation area intended to develop was 15 000 ha.

The IGME (1982) model was a modified version of the finite differences USGS (Pinder-TRescott-Larson) code. The nodes of its rectangular grid were variable in size. Ocean contact was assumed a constant-head boundary. Connections between rivers and lakes and the aquifer were included. Evapotranspiration was dependent on water table depth. The later IGME (1987) model was apparently run with the same grid, boundaries and code. Only the pumpage and the natural recharge were slightly changed.

In order to contribute to a better knowledge of the critical zone (ecotone and "cotos" near El Rocío) a more detailed study was done (SUSO 1988, SUSO & LLAMAS 1989). Figures 5

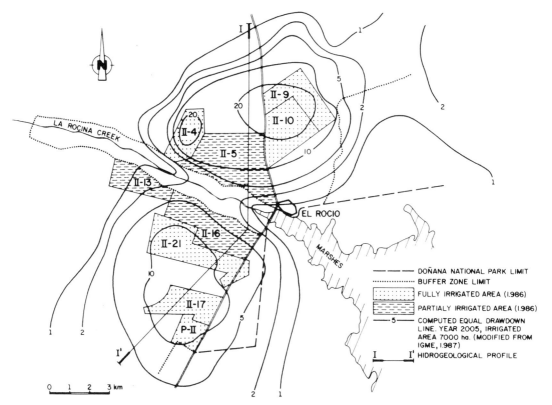

Fig. 5: *Computed watertable drawdowns at the El Rocío zone after 20 years of pumping (52 hm³/year for irrigation of 7000 ha after IGME (1987) numerical model (after SUSO & LLAMAS 1989)).*

and 6 show some significant results from this work. Nevertheless, the irrigation and touristic development of the area continued. Currently (1989) the irrigated surface is about 10 000 ha.

In november 1988 a Mission sent by the International Union for Conservation of Nature (IUCN) and by the Worl Wide Fund (WWF) visited the area (LLAMAS 1988b). The English version of its Report thereon (HOLLIS et al. 1989) has been distributed in July 1989. This Report definitely supports the concerns about the serious environmental impact caused by the groundwater extraction from the lower Giadalquivir valley aquifer.

9 Conclusion

Frequently, ecosystems located on eolian sands are dependent on the groundwater flow system inside and outside those eolian sands. Such a flow system can only be properly understood if it is studied within a broader hydrogeological frame.

In countries where a strong demand for groundwater exists it seems crucial to have a thorough knowledge of the hydrological functioning of the ecosystems

Geohydrology of Eolian Sands, Spain

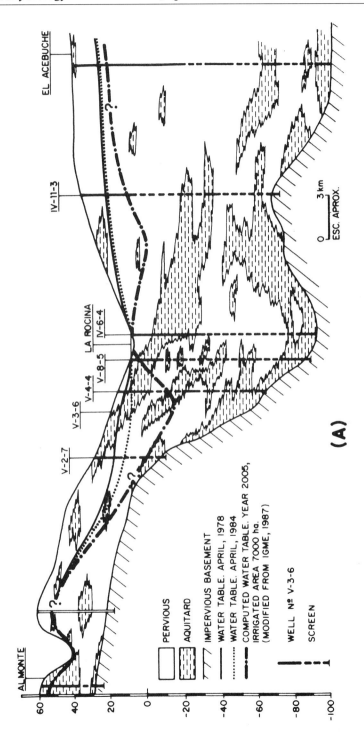

Fig. 6: *Hydrogeological profile I-I. Situation in fig. 5 (after SUSO & LLAMAS 1989).*

located on eolian sands if these ecosystems are to be protected. At the same time a more decisive and timely action is required in terms of the hydrogeological education of the Government officials involved, of the politicians and of the general public; and also in terms of finding financial compensation for these who are adversely affected by this conservation policy.

References

ALLIER, C., GONZALEZ BERNALDEZ, F. & RAMIREZ, I. (1974): Mapa ecologíco de la Reserva Biológica de Doñana, Escala 1:10 000. C.S.I.C. Seville, Spain.

HOLLIS, T., HEURTEAUX, P. & MERCER, J. (1989): The implication of Groundwater Extractions for the long Term Future of the Doñana National Park. Report of the WWF/IUCN/ADENA Mission to the Doñana National Park, May 1989, 60 pp.

LLAMAS, M.R. (1988a): Conflicts between Wtland Conservation and Groundwater Exploitation: Two Case Histories in Spain. Environmental Geology, Vol. 11, No. 3.

LLAMAS, M.R. (1988b): Analysis of the impact of groundwater exploitation on the Doñana ecosystems. Remarks for the WWF/IUCN Commission, Report for the WWF Mission, November 1988, 9 pp.

LLAMAS, M.R. (1989a): The Role of Groundwater in the Functioning of Mediterranean Wetlands. Presentation at the EEC Wetlands meeting at Exeter University (U.K.), March 1989 (at press).

LLAMAS, M.R. (1989b): Wetlands and Groundwater: New Constraints in Groundwater Management. Intern. Assoc. of Hydrological Sciences, Publ. No. 188, 595–604.

LLAMAS, M.R., RODRIGUEZ, J., TENAJAS, J. & VELA, A. (1987): El Parque Nacional de Doñana: el medio físico. Seminario sobre Bases Científicas para la protección de los humedales en España, Real Academia de Ciencias, Madrid, 209–216.

MERINO, O. & MERINO, Y. (1988): El impacto potencial de la explotación del acuífero Almonte-Marismas en los ecosistemas del área de Doñana. Intern. Symp. on Hydrology of Wetlands in Arid and Semi-arid Zones, Agencia del Medio Ambiente, Seville, Spain. 123–128.

RODRIGUEZ AREVALO, J. & LLAMAS, M.R. (1986): Groundwater development and water table variation in the Doñana National Park. Memoirs of the 19th Congress of the International Association of Hydrogeologists, Karlovy-Vary, 203–211.

SUSO, J.M. (1988): Estudio hidrogeológico de la influencia de la extracción de aguas subterráneas en las proximidades de El Rocío (Huelva). Master thesis, Faculty of Geology, Complutense University, Madrid.

SUSO, J.M. & LLAMAS, M.R. (1989): Impact of groundwater exploitation on wetlands: Doñana National Park (Spain) case. Abstract 28th Intern. Geological Congress, Washington, pp.3-188-89.

VELA, A. & LLAMAS, M.R. (1986): Análisis preliminar del flujo del agua subterránea en el sistema de dunas móviles del Parque Nacional de Doñana. Actas del II Simposio sobre el Aqua en Andalucía, University of Granada, Vol. 2, 423–434.

Address of author:
M. Ramón Llamas
Dept. of Geodynamics
Complutense University
28040 Madrid
Spain

THE CHARACTERISTICS OF DUNE SOILS

P.D. **Jungerius**, Amsterdam

Summary

Dune soil profiles express the relative importance of geomorphological and biological processes. Depending on the balance between these processes, three compartments can be recognized. These compartments are arranged in a zonal sequence as well as, on a smaller scale, a catenary sequence.

1 Introduction

Literature on soils in coastal dune areas is not abundant. From the middle of the last century onwards, soil science developed because it supported the rapidly expanding agricultural production. As dunes are hardly important from an agricultural point of view, little attention has been given to their soils. In most national soil classifications dune soils are lumped into an undifferentiated and poorly described group, a fate they share with mountain soils and other soils of 'wild' areas. However, their restricted economic value does not mean that they are pedologically uninteresting. On the contrary, compared to soils under agriculture which are largely fossil below the manmade A horizon, dune soils offer the opportunity to study several interesting pedological processes 'in vivo'.

To make the objective of this study clear, it is useful to consider its position in the classification of soil studies proposed by BUTLER (1958). BUTLER identified three concepts about soils: the edaphic concept, the pedologic concept and the geographic concept. Up to now, dune soils have been considered mainly from the **edaphic** viewpoint by ecologists. Their field of interest is restricted to soil conditions which are of significance to the growth and life economy of plants. Ecologists and botanists have early realized the importance of dune soils as a medium for the growth of plants. However, soil properties considered relevant to plant growth are often limited to chemical variables such as pH, $CaCO_3$ content and organic matter content (SALISBURY 1925, BOERBOOM 1963, RANWELL 1972).

The **pedologic** concept comprises objective studies of the development of the soil profiles and soil materials, the principles of their occurrence and the processes of their origin such as desalinization, decalcification, podzolization and formation of organic matter. Purely pedologic studies of dune soils are rare (KLIJN 1981).

Although influenced by the other concepts, the present chapter is written with the **geographic** concept in mind. In this concept, a soil is a dynamic three dimensional piece of landscape. This has been for many years the guiding princi-

ISSN 0722-0723
ISBN 3-923381-23-9
©1990 by CATENA VERLAG,
D-3302 Cremlingen-Destedt, W. Germany
3-923381-23-9/90/5011851/US$ 2.00 + 0.25

ple of the Netherlands' Soil Survey for many of their surveys including those of coastal areas (STIBOKA 1967a & 1967b, KLOOSTERHUIS 1980, MARKUS & VAN WALLENBURG 1982, VOS 1984, VAN OOSTEN & KUYER 1986).

In dealing with the soils in the dunes, an ecological line will be adopted for this chapter. This line issues from the interest in soils as part of a natural landscape. This has focussed attention on the close relationship of soil dynamics with biotic and geomorphic processes of the site. From this it follows that relief and vegetation are considered to be the most important soil forming factors. Their relationship with dune soils will be discussed first.

For soils formed under natural conditions KLINKA et al. (1981) have developed a system of soil profile description which pays due attention to the organic component of the soil. Their system is referred to in this chapter because it is better suited to soil studies in the dunes (WARDENAAR & SEVINK 1990) than the standard FAO/UNESCO Guidelines (1975).

2 Dune relief as a soil forming factor

The topography of the dunes influences soil profile development in many ways. Dune forms control wind regimes (FRYBERGER & DEAN 1979) and thereby determine the pattern of wind erosion and accumulation. Water erosion affecting dune soil profiles is a function of slope and exposition (BRIDGE & ROSS 1983, RUTIN 1983). Relief exerts also indirect control through regulation of (micro-)climate, vegetation and groundwater level. For example, compared to their counterparts on the southern side, north exposed slopes have lower temperatures, less insolation and a lower evaporation rate. Their vegetation cover is generally denser and more luxuriant. As a consequence, living conditions for the soil fauna are much better than on south exposed slopes. Organic matter is of different type (MÜCHER this volume), organic matter content is also higher and soil horizons are better expressed (see also GERLACH 1989).

Groundwater affects soil profiles only in dune valleys and in seepage areas at the landward boundary. Under permanently waterlogged conditions the familiar grey reduction colours are found. In the zone with seasonally or episodically fluctuating watertable gley mottling is produced. Groundwater controls soil profile formation also through plant growth and mineralization rates.

An example of the relationship between various topographic variables and soil profile properties derived from an actual survey of dry dunes (DUIJN 1988) is given in tab. 1. The area is located north of Haarlem on the coast of the Dutch mainland (Appendix). The first two rows in this table represent situations where the soil surface has been covered by windblown or colluvial sand. In the soils of the first row the addition has occurred regularly and either slowly (AC horizon at the surface) or rapidly (C horizon at the surface). These soils are frequently found in higher terrain with southeast exposition (tab. 1). The soils of the second row are covered discontinuously but in massive amounts, resulting in the complete burial of the surface horizon. This is characteristic for flat or nearly flat bottom situations.

The remaining rows picture soils in the order of increasing stability. In the third

	n	S1	S2	S3	O1	O2	O3	H1	H2	H3
AC or C horizon at surface	15	40	13	47	7	27	66	13	47	40
buried A horizon in profile	21	76	14	10	38	42	20	52	24	24
solum > 20 cm thick	64	61	23	16	35	26	39	41	42	17
organic horizon at surface	23	74	13	13	52	13	35	61	30	9
B horizon present	19	74	26	0	63	16	21	53	47	0

S = Slope angle: S1 0–5°; S2 5–15°; S3 >15°
O = Orientation: O1 none; O2 315–135° (NE); O3 135–315° (SE)
H = Height: H1 0–5 m; H2 5–15 m; H3 >15 m

Tab. 1: *The relationship between topographic variables and soil profile properties in the dune north of Haarlem (from DUIJN 1988). Frequencies are shown in horizontal percentage per group of topographic variables.*

row soil development has advanced sufficiently to include an A horizon of more than 20 cm. Such soils are often found on the lower parts of slopes which are not steep and without preferential orientation. In the next phase, represented by row 4 of tab. 1, at least one of the organic horizons, mostly L and F, has been identified. Such soils have a strong preference for low-lying level sites. This also applies to soils in which profile development has progressed to the stage that a brown B horizon is included in the solum (see MÜCHER this volume). It is clear from this table that soil profiles are better developed with decreasing slope angle and decreasing altitude, which indicates the importance of the geomorphological dynamics.

3 Vegetation as a soil forming factor in the dunes

The plants furnish the organic matter content of the surface soil, but the actual quantity produced depends on the balance between the rate of biomass production and the rate of breakdown. Limiting factors for the decomposition and mineralization are lack of oxygen and low temperatures or low pH levels. The first condition is found in waterlogged soils in dune slacks, the latter in leached, rather acid profiles.

The study of the relationships between vegetation and dune soils is complicated by a number of factors. Each soil type can support a range of species with very different edaphic requirements. Clearly the soil is but one of the environmental aspects which have ecological relevance for plant growth in the dunes. Also, very little is known of the strategies used by plants to suit their environment which includes the soil to their purpose. Although fruitful efforts have been made to associate vegetation, succession and soil profile development (e.g. SALISBURY 1925, BOERBOOM 1958, RANWELL 1972), more detailed studies of microhabitats are needed. One promising alley

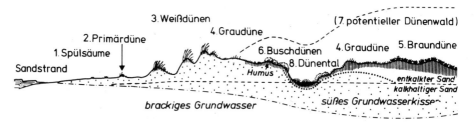

Fig. 1: *Cross section of a coastal dune area along the Northsea (ELLENBERGER 1978, 493).*

vegetation type:	mosses + lichens	grassland	shrubland Hippophaë	shrubland Salix	shrubland Crataegus
n:	36	41	15	10	8
soil property: depth F + A hor. (cm)	3.3	6.8	4.1	5.3	11.4
F horizon present (%)	6	29	20	20	63
buried A hor. present (%)	61	29	20	60	30
depth decalcification (cm)	3.8	7.2	3.9	5.7	16
soil fauna present (%)	28	73	73	100	100
root index	43	90	64	78	123

Tab. 2: *The relationship between vegetation types and soil profile properties in Meijendel near The Hague (from SMITS 1989). The root index is a relative measure obtained by multiplying the depth of occurrence with the abundance class number of very fine and fine roots on the scale of the FAO/UNESCO Guidelines (1975).*

is provided by micromorphological analysis (MÜCHER this volume). With this technique it is possible to subdivide soil organic matter and $CaCO_3$ particles into components of different origin and with different edaphic properties.

Recently a number of soil surveys has been carried in the Dutch dunes based on the mapping units of existing vegetation or landscape-ecological maps. Tab. 2 depicts soil properties encountered in the various mapping units of the landscape-ecological map of the dunes near The Hague (VAN DER MEULEN & VAN HUIS 1985) by SMITS (1990). The surveyed area is characteristic for the grey dunes (fig. 1).

1. Pioneer vegetation with grey hairgrass (*Corynephorus canescens*) and wall pepper (*Sedum acre*), and abundant moss and lichen (tab. 2, column 2).

Vegetation cover is often incomplete (up to ca. 15% bare sand) and geomorphic dynamics are high. Organic horizons, if present, consist of the fermentation layer (F horizon). A horizons are poorly developed with few roots and little activity of the soil fauna. Buried A horizons are common on sites with frequent sedimentation of windblown sand or colluvium.

2. Dune grasslands with sheep sorrel (*Rumex acetosella*), red fescue (*Festuca rubra*), wood small-reed (*Calamagrostis epigejos*), dewberry (*Rubus caesius*) and restharrow (*Ononis repens*) (tab. 2, column 3).

Decalcification has progressed, the F and A horizons are much more expressed, buried soils are scarce, and plant root systems are well developed.

3. Shrubland dominated by sea buckthorn (*Hippophaë rhamnoides*), creeping willow (*Salix repens*) and hawthorn (*Crataegus monogyna*), respectively (tab. 2, Columns 4, 5 and 6).

In this direction the depth of the F and A horizons increases, as does the depth of decalcification. As time proceeds, decalcification causes lower pH values, decreasing availability of nutrients, and the development of humus types with lower C:N ratios. The variability of the microclimate decreases with increasing density of the stands. Influence of the soil fauna is found in most profiles. Buried soils are most frequent under *Salix* which is best adapted to being overblown by sand. Root systems are best developed in hawthorn shrubland which occupies the more humid sites.

4 The formation of soil as part of the dune landscape

In this section the concept proposed by JUNGERIUS & VAN DER MEULEN (1988) will be followed. This concept emphasizes the importance of interruptions of pedological processes by the action of wind and water, and leads to a clearly recognized zoning of the landscape of the dry dunes.

The landscape forming processes relevant for soil formation in the dunes can be divided into two groups. The first deals with geomorphological processes which represent the unstable factor of landscape development and leads to rejuvenation of the soil profile. The second group, comprising the biological processes, generally encourage the stabilization of the dune terrain and the formation of well developed soil profiles.

Depending on the balance of the opposing geomorphological and biological processes, three compartments can be recognized in a dune landscape. These compartments are arranged in a zonal sequence, as well as, on a more detailed scale, a catenary sequence.

4.1 Zonal sequence

4.1.1 The soils of compartment I

In the first compartment no organic or A horizons are formed because the geomorphological processes are too active and shift large masses of sand. Influx of NaCl by seaspray is usually high. Some plant species are able to withstand the extreme geomorphological and climatological activity, but there is usually not much more than a scarce pioneer vegetation (*Ammophila, Carex, Elymus*). This is the zone of the 'yellow' dunes which comprises the foredunes and adjacent zones where sand is in continuous movement or is recently fixed (fig. 1).

4.1.2 The soils of compartment II

The second compartment is characterized by great variability in the balance between the two groups of processes. It is the zone if 'grey' dunes which forms a large part of the inner dunes. The variability applies to space as well as to time. Places which are completely stabilized by vegetation alternate with places where the soil surface is bare due to erosion or sedimentation. As a result there is also great variability in the development of the soil profile (fig. 2).

On stable sites the development of a well defined B horizon is no exception, but shallow and truncated soils prevail where erosion is an active process, especially on upper parts of slopes steeper than 15°. The profiles lower down on these slopes reflect the balance between colluviation and soil formation. Where colluviation is so slow that soil formation can keep pace with the accumulation of sand, an abnormally thick Al horizon is formed. Conversely, if colluviation is a rapid process there is hardly an Al horizon discernible because the surface is covered before organic matter can accumulate.

Temporal variability is as common as spatial variation. Sites where periods of colluviation alternate with periods of plant growth and stability are characterized by a sequence of buried A horizons. On other than lower slope and bottom situations, a similar sequence of profiles can be found where aeolian sand is deposited on vegetated surfaces.

4.1.3 The soils of compartment III

In the third compartment geomorphological processes are no longer important and the surface is stabilized by plant growth. The natural vegetation type is often characterized by forest with deciduous trees (*Quercus, Betula, Populus*). Soil development can continue uninterruptedly. The relatively well developed soil profiles of compartment III are found in the densely vegetated 'brown' dunes farthest from the sea (fig. 1), but also in protected areas of the two other compartments.

Probably the best-known soils in wooded dune are the podzols of Les Landes in SW France (see BRESSOLIER, this volume). The induration of the illuvial horizon into a hardpan (alios) after removal of trees interfered with the reestablishment of trees or the growth of other crops. In many other dune areas along the European coast, the large scale introduction of conifer species has led to the formation of soils which differ strongly from soils formed under the natural vegetation. WARDENAAR & SEVINK (1990) compared two adjacent sites with forest stands of about the same age (80 years), one being a planted *Pinus sylvestris* stand, the other a naturally established *Populus nigra* L. stand. The parent material of both plots is similar, groundwater occurs at great depth and climatic differences are negligible.

Soils in coniferous forest exhibit a well developed organic profile and a more or less prominent micropodzol, indicating a low decomposition rate and minimal bioturbation. The organic matter is of the mor type, poor in nutrients and with a low pH, in spite of the calcareous sand at shallow depth. In contrast, the decomposition rate of the organic matter in

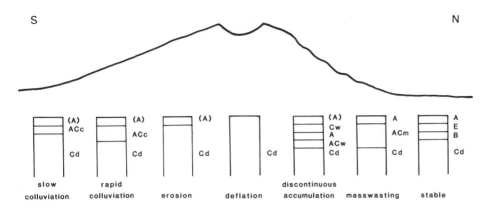

Fig. 2: *Characteristic soil profiles of a ridge with an active blowout in the grey dunes compartment.*

soils of deciduous forest is much higher and the organic horizons are relatively insignificant. Bioturbation is more important than in soils of coniferous forest. Consequently, the A horizon is thick and the boundary between the horizons is gradual. Organic matter is of the mull type and the C/N ratio is much lower than under coniferous trees.

4.2 Catenary sequence

On a smaller scale within a compartment, even on one and the same slope, there are often occurrences of the other compartments: a ridge in the grey dunes (compartment II) may have an active blowout in the summit where fresh C material is exposed (compartment I), and at its base a dense woodland which may have existed long enough for the soil to show a distinct profile differentiation (compartment III).

Whatever the scale the dune landscape is looked upon, the soil profile expresses the relative importance of geomorphological and biological activity. The soil profile is therefore a powerful source of information on the dynamics of the dune landscape.

5 Man as a soil forming factor in the dunes

Althoug less destructive than in agricultural areas, man has exerted his influence on dune soils as long as these soils exist. His activities have direct as well as indirect consequences. To the direct consequences belong the destruction of the surface horizon by trampling in recreational areas or by removing the sod to suppress the growth of unwanted plant species in nature reserves. Many afforestation schemes involved thorough reworking of the surface soil.

Of indirect consequences are changes effectuated in any one of the soil forming factors, from climate in the case of acid rain and the greenhouse effect down to vegetation in the case of planting marram grass or cutting trees. Space does not permit to discuss these problems into detail but it is clear they add to the vast amount of subjects which are left to be investigated, in order to elucidate the role of the soil in the dune ecosystem.

References

BOERBOOM, J.H.A. (1958): Begroeiing en landschap van de duinen onder Scheveningen en Wassenaar van omstreeks 1300 tot heden. In: Adviescommissie Duinbeplanting, Beplanting en Recreactie in de Haagse duinen. Meded. ITBON 39, 108 p., Wageningen.

BOERBOOM, J.H.A. (1963): Het verband tussen bodem en vegetatie in de Wassenaarse duinen. Boor en Spade, 120–155.

BRIDGE, B.J. & ROSS, P.J. (1983): Water erosion in vegetated sand dunes at Cooloola, southeast Queensland. Z. Geomorph. N.F. Suppl.-Bd. 45, 227–244.

BUTLER, B.E. (1958): The diversity of concepts about soils. J. Austr. Inst. Agric. Sci. 14–20.

DUIJN, R. (1988): Verslag over de relaties tussen bodem, vegetatie en fysiografische ligging in het Noordhollands Duinreservaat. Rapport Fysisch Geografisch en Bodemkundig Laboratorium en Provinciaal Waterleidingbedrijf van Noord-Holland. 65 p.

ELLENBERG, H. (1978): Vegetation Mitteleuropas mit den Alpen. Verlag Eugen Ulmer, Stuttgart, 981 p.

FAO/UNESCO (1975): Guidelines for soil profile description. Rome, 53 p.

FRYBERGER, S.G. & DEAN, G. (1979): Dune forms and wind regime. In: McKee (Ed.), A study of global sand seas. U.S.G.S. Prof. Paper 1052, 137–170.

GERLACH, A. (1989): Biochemistry of nitrogen in coastal dune ecosystems. Paper submitted to the European Conference on Landscape Ecological Impact of Climatic Change, Lunteren, The Netherlands.

JUNGERIUS, P.D. & VAN DER MEULEN, F. (1988): Erosion processes in a dune landscape along the Dutch coast. CATENA 15, 3/4, 217–228.

KLIJN, J.A. (1981): Nederlandse kustduinen; geomorfologie en bodems. Pudoc, Wageningen, 188 p.

KLINKA, K., GREEN, R.N., TROWBRIDGE, R.L. & LOWE, L.E. (1981): Taxonomic classification of humus forms in ecosystems of British Colombia. Min. of Forests, Canada.

KLOOSTERHUIS, J.L. (1986): Blad Texel. Netherlands Soil Survey, Wageningen.

MARKUS, W.C. & VAN WALLENBURG, C. (1982): Blad 30-W en 30-O. Netherlands Soil Survey, Wageningen.

RANWELL, D.S. (1972): Ecology of salt marches and sand dunes. Chapman & Hall, London, 258 p.

RUTIN, J. (1983): Erosional processes on a coastal sand dune, De Blink, Noordwijkerhout. Publicaties van het Fysisch Geografisch en bodemkundig Lab. 35, 144 p.

SALISBURY, E.J. (1925): Note on the edaphic succession in some dune soils with special reference to the time factor. J. Ecol. 13, 322–328.

SMITS, P. (1990): Duinbodems en hun vegetatie. Rapport Duinwaterleiding van 's-Gravenhage en het Fysisch Geografisch en Bodemkundig Laboratorium van de Universiteit van Amsterdam.

STIBOKA (1967a): Blad 36, Goedereede. Netherlands Soil Survey, Wageningen.

STIBOKA (1967b): Blad 42-O, Zierikzee. Netherlands Soil Survey, Wageningen.

VAN DER MEULEN, F. & VAN HUIS, J.C. (1985): Dune landscape map of Meijendel. Duinwaterleiding van 's-Gravenhage.

VAN OOSTEN, M.F. & KUYER, P.C. (1986): Kaartbladen 1, 2 en 4, Vlieland, Texel, Ameland, Schiermonnikoog. Netherlands Soil Survey, Wageningen.

VOS, G.A. (1984): Blad 37-W, Rotterdam. Netherlands Soil Survey, Wageningen.

WARDENAAR, C.P. & SEVINK, J. (1990): A comparative study of soils under first generation Scots Pine and Poplar stands on calcareous coastal sands. In preparation.

Address of author:
P.D. Jungerius
Landscape and Environmental Research Group
Univ. of Amsterdam
Dapperstraat 115
1093 BS Amsterdam
The Netherlands

MICROMORPHOLOGY OF DUNE SANDS AND SOILS

H.J. Mücher, Amsterdam

Summary

Pedological and micromorphological techniques have been applied to the study of abiotic (lithology, relief, groundwater and soil) and biotic (vegetation and fauna) landscape-forming factors. Examples are given of the preliminary results of recent micromorphological studies of the dunes near Castricum, The Netherlands.

Future fields of micromorphological and pedological research in the dune area are:

- Compositional and structural analysis of thin sections to contribute to the characterization and history of the sands, to their mode of formation and postdepositional changes.

- The study of the pedological features of the various vegetation communities in order to understand the interaction between plant and soil and explain vegetation succession.

- Paleogeomorphological studies of the dynamics of the dunes as reflected in the occurrence of paleosols.

1 Introduction

During the last decade the authorities' interest in coastal dune areas has grown (VAN DER MEULEN et al. 1989), and the role of the dune managers has become more and more important. This development demands more knowledge of the individual factors controlling the dune landscape, and the interaction between these factors. These factors are: climate, lithology, relief, groundwater, soil, vegetation and fauna (BAKKER et al. 1979).

The aim of this article is to demonstrate how pedology in combination with micromorphology (i.e. microscopic study of undisturbed soil samples in thin sections) can contribute to a better understanding of the above mentioned factors and their interactions.

2 Abiotic factors: lithology, relief, groundwater and soil

2.1 Lithology

The grain-size of dune sands is rather uniform, lying mainly between 105 and 300 microns in diameter. The minerals are dominantly quartz, and locally they occur in combination with carbonates. Scattered grains of heavy minerals and rounded quartzite or sandstone fragments occur. South of Bergen, the Dutch coastal dunes are mainly calcareous and north of this township the carbonate content is smaller. This is primarily related to the source area of the sand and secondly to decalcification during the Pleistocene and early Holocene periods, when the North Sea Basin was not flooded.

Hitherto attention has been mostly confined to the macro-nutrients and pH of the dune sands, in relation to plant growth (BOERBOOM 1963). Little is known, however, of the morphology of the carbonate components and their mineralogical composition. A micromorphological study of ASSENDORP & MÜCHER (1990) in the dune area of Meijendel near The Hague, showed that in thin section three types of carbonates could be recognized in the soil horizons:

- Shell and shell fragments

- Primary carbonate crystals

- Nodules of secondary carbonates (micrite). The nodules are mostly of the same size and roundness as the quartz grains. This morphology suggests that the carbonate nodules have been deposited together with the other mineral components.

Accumulations of secondary carbonates formed in situ (photo 1) are almost never observed. Secondary carbonates mainly occur very locally in channels near root remnants. On the other hand, solution rims are observed at the surface of all of the three carbonate types. Identification of the mineral composition of the various carbonates is only possible by chemical tests. Making use of polished impregnated undisturbed samples, the various carbonate minerals can be identified by their change in colour.

Discontinuous, patchy, and very thin (<20 microns), coatings of organic material, clay and iron, occur at the surface of the sand grains. The organic coatings are best developed in the A horizons (photo 2), whereas the patchy cutans of clay and iron are well developed in colour B horizons. The water repellent behaviour of dry greyish sands is probably caused by the presence of certain types of organic coating (DEKKER & JUNGERIUS, this volume). It is suggested here that these coatings are chemical decomposition residues of organic material, and not of faunal decomposition of organic residues, e.g. residues of aged organic excrements. This is in agreement with the occurrence of water repellent sand mainly on south facing slopes, where faunal biological activity is much lower than on slopes north facing (see also par. 3.2: Fauna). Our micromorphological observations suggest that water infiltrates downwards until a water repellent spot on a sand grain has been reached.

2.2 Relief

The dune ecosystem is characterized by the interaction of geomorphological and biological processes. The geomorphological processes are mainly the action of wind and water. The biological processes comprise development of vegetation, production of soil organic matter and faunal activity, resulting in soil profile differentiation.

Discrimination of sandy materials redeposited by water or by wind is not usually possible. A valuable contribution to geomorphology would be a micromorphological distinction between the structures of translocated sand by wind and water.

Micromorphology of Dune Sands and Soils

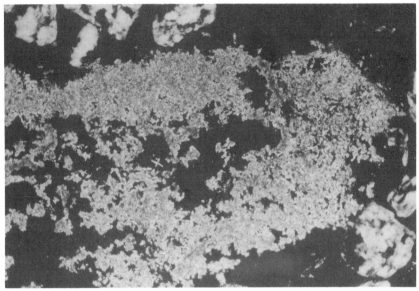

Photo 1: *Microphotograph of a thin section showing secondary carbonate precipitation in a void of a greyish brown (10YR 5/2) AE horizon at a depth of ca. 6 cm. The AE horizon occurs in a dune slope with SE exposition and an open moss cover of 10% in the Meijendel area, The Netherlands. Frame length: 1.3 mm.*

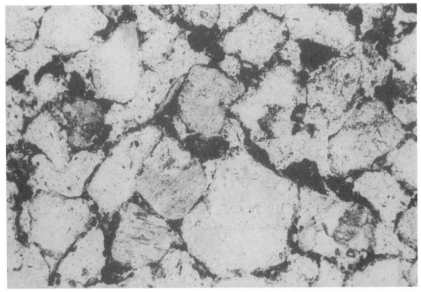

Photo 2: *Microphotograph of a thin section showing organans around dune sand grains and incomplete fillings of organic material in between the sand grains. Located in an Ab horizon with greyish humic sand (10YR 5/1-4/1) at a depth of ca. 14 cm in a dune slope in the Meijendel area, The Netherlands. Frame length: 1.3 mm.*

2.3 Groundwater

Hydromorphic features, such as greyish zones of reduction, and mottles or cutans of iron and manganese (hydr)oxides, caused by recent or former groundwater tables can be recognized in thin section. To distinguish between recent and paleohydromorphic phenomena is not always possible. Active hydromorphism can be recognized as such if it occurs in relation to recent biopores and/or the presence of a nearby water-table. POELMAN et al. (1989) demonstrate that periodic saturation of poorly drained sandy soils in The Netherlands results not only in a mottled soil, showing mottles with chroma's of 2 or less, but also in significantly higher bulk densities of mottled A, B and C horizons with organic contents of smaller than 2.5%, than those of the corresponding unmottled horizons. This is in agreement with the variation in penetration resistance measurements of BAKKER et al. (1979) in the parched dunes of Schoorl, The Netherlands, resulting from periodic groundwater fluctuations. These higher bulk densities restrict rootpenetration and result in different hydraulic conductivity and moisture retention characteristics, as compared with unmottled horizons. Thus, after the lowering of water tables by natural causes or by water management practices, some physical effects of the previous period of saturation may persist, adversely influencing plant growth. This is a promising avenue for further micromorphological research in the dunes.

2.4 Time

The "Dutch Young" dune landscape was formed in the 10th century (KLIJN, this volume), since when it has been modified by erosion and deposition. From the pedological point of view the dunes are young to very young, although decalcification has affected the parent material in many places.

The short period of soil formation is reflected in a weak soil formation, giving rise to a 0-Ah-C, and occasionally to a 0-Ah-B-C horizons development. The brown colour of the B horizon is the result of iron coatings around the grains, or is due to fulvic acids washed in by rainwater and covering the sand grains (SLIKKER & STOKVIS 1977). Micromorphological observations of ASSENDORP & MÜCHER (1990) revealed that grains with organic patchy coatings, as well as grains with patchy cutans of iron and clay occurred in all soil horizons, ranging from 10–30% and 10–20% by volume respectively (see also previous paragraph 2.1: Lithology).

From the micromorphological point of view too little attention has been given to the characteristics of the soil surface in dune areas, in relation to the distribution of certain plant species. For example, a study of MÜCHER et al. (1988) in semiarid rangeland areas of Australia suggests that the soil surface conditions affect the lodgement and germination of seeds, and the infiltration and runoff of water, thereby influencing plant growth. Subsequently CHARTRES & MÜCHER (1989) experimentally demonstrated the effect of the surface conditions of the soil on the entrapment and germination of wallaby grass seeds (*Danthonia caespitosa*; synonym *Rytidosperma caespitosum*).

3 Biotic factors: vegetation and fauna

3.1 Vegetation

The development of certain vegetation communities is determined by a complex interaction of various factors, such as: climate, micro climate (exposition, relief, distance from the shore line, soil temperature, evaporation), parent material, soil development (inclusive depth of decalcification), groundwater level, and human activity, such as reclaiming land and changing the groundwater level (SLIKKER & STOKVIS 1979, BOERBOOM 1963, ZONNEVELD 1959). From the micromorphological point of view the study of the site characteristics necessary for the development of a certain vegetation community, and of the influences of a certain plant association at a location, is still in its infancy.

3.2 Fauna

On one hand animals, for example rabbits, reduce the vegetation cover by grazing and dig up sand, thus promoting erosion of the slope, on the other hand various species of the soil fauna produce organic matter that gives rise to the growth of fungi hyphae and bridges of humic substances that stabilize the sand grains (BARRATT 1962, JUNGERIUS & VAN DER MEULEN 1988).

An interesting review paper by HOLE (1981) considers the effects of animals on soil. Although the fauna species are only incidentally observed in thin sections (they disappear during sampling, transport and drying of the undisturbed soil samples), some results of their activities can be studied in thin section, such as: biopores, bioturbation backfilling voids, and excrements. The excrements are important micromorphological features, which reflect former or current animal activity and environmental conditions. In addition they often form an essential part of the soil structure (BULLOCK et al. 1985). There are, however, few publications concerning the micromorphology of excrements. JONGERIUS (1957), ZACHARIAE (1965), BAL (1973 & 1982) and RUSEK (1975) consider excrements in terrestrial soils (only occasionally of cover sand or drift sand areas), and KOOISTRA (1978) investigated micromorphologically excrements from an intertidal zone in the Oosterschelde area, southwestern Netherlands.

Micromorphological analysis of north and south facing dune slopes in the coastal dune area Meijendel near The Hague (The Netherlands) were recently made by the author. The preliminary results suggest that the faunal activity is much larger on the north facing slopes than on those facing south. This results in an important production of organic excrements of the moder type on the north facing slopes, whereas on the south facing slopes the litter and root remnants were largely amorphous organic substances, probably modified by micro biological and chemical processes. Soil animals that could be considered responsible for the production of these moder fecal pellets are, according to JONGERIUS (1957): *Oribatei* (mites), *Collembola* (spring tails), *Isopoda* (armadillo bugs, pill bugs and wood lice), and *Enchytraeidae* (pot worms).

4 Micromorphological case studies in the dunes near Castricum ("Noordhollands Duinreservaat"), The Netherlands

In the summer of 1988 two studies were made by DUIJN (1988) and VAN DER ZIJP (1989), with the aim of characterizing the micromorphological and pedological features of the various vegetation communities as mapped by KRUYSEN (1990). The following examples are taken from these studies. The area of study is located in more or less stabilized calcareous dunes, at moderate distance from the sea, where landscape and soils are characterized by the interaction of geomorphological and biological processes (JUNGERIUS & VAN DER MEULEN 1988). The five vegetation communities described below have been arranged in the order of decreasing geomorphological and increasing biological dynamics:

- Geomorphological activity is greatest in vegetation community 12K. The vegetation cover is incomplete and the plant species are resistant to frequent accumulation of windblown or colluvial sand.

- Both geomorphological and biological dynamics are moderate in the dune grassland communities 13K and 3K, the difference being the occurrence of buried A horizons in the latter community which indicates the occasional occurrence of unstable periods of windblown sand accumulation.

- The moss-covered sites of vegetation community 13m are also characterized by buried soils, but the interruptions by unstable periods seem to have ended, and the present soil is better expressed.

- Vegetation community 54 is found in slacks and places with slopes of low angle and little geomorphological activity. Vegetation cover is dense, with tall herbs and scattered trees or bushes, and soils here are relatively well developed.

Vegetation community 12K: *Ammophila arenaria, Calamagrostis epigeos* and *Rubus caesius* (70%) with herbs.
Site characteristics: Elevation: 13 m; lower slope (1°) near valley bottom; exposition: SW; distance to shore line: 200 m; vegetation cover: 75–100%; soil horizons: AC-C (soil depth >11 cm).
Soil characteristics: The soil profile of vegetation community 12K is weakly developed, showing a low biological activity (e.g. few plant remains with few excrements and few pedotubules). The shell fragments exhibit solution phenomena near the rims only in the AC horizon. The soil profile reflects a low activity of biological processes and a high activity of geomorphological processes (colluviation).

Vegetation community 13K: dune grassland with *Rubus caesius*, herbs, mosses and lichens.
Site characteristics: Elevation: 11 m; lower slope (1°) near valley bottom; exposition: W-NW; distance to shore line: 150 m; vegetation cover: 75–100%; soil horizons A-AC-C (total thickness >56 cm).
Soil characteristics: The soil profile below community 13K shows a low biological activity in all horizons (with exception of the upper A horizon). The organic material is mainly chemically

transformed in organic irregular aggregates, and only partly composed of welded fecal pellets. Shell fragments are absent from the A horizon, and occur only in the deeper soil horizons in combination with carbonate nodules. Most of the shell fragments and nodules show solution rims. Pedotubules occur only in the upper part of the A horizon.

Vegetation community 3K: dune grassland with *Rubus caesius*, many herbs, and mosses (25%).

Site characteristics: Elevation: 6 m; upper slope (18°) at the edge of a blowout; exposition: N; distance to shore line: 140 m; vegetation cover: 75%; soil horizons: A-C-2Ab-2ACb-2C (total soil depth >21 cm).

Soil characteristics: The recent soil on top shows a weakly developed A horizon with common litter and root remnants, many excrements and no pedotubules. This soil contains few shell fragments and carbonate nodules of micrite. The nodules have almost disappeared from the uppermost A horizon. The buried soil profile is more strongly developed resulting from the increased biological activity which is demonstrated by the occurrence of abundant excrements, the few plant remnants, and more, but still few, pedotubules in the buried A horizon than in the top soil. The A horizon contains fewer micritic nodules than the C horizon, which suggests that the carbonate nodules have been dissolved by organic acids. The soil development reflects an alternation of geomorphological and biological processes (soil formation).

Vegetation community 13m: *Hypnum cupressiforme* with winter annuals, grasses and lichens.

Site characteristics: Elevation: 4.5 m; flat dune valley bottom; exposition: -; distance to shore line: 960 m; vegetation cover: 90%; soil horizons: Ah-AC-C-2Apb-2Cb-3Ahb-3Cb (total soil depth >52 cm).

Soil characteristics: The soil profile of vegetation community 13m has two buried soils which contain no shell fragments. The biological activity is relatively high in all the soil horizons, as shown by the common occurrence of plant and root fragments, pedotubules and excrements. Discontinuous, very thin cutans of clay mixed with iron, or of organic material around the sand grains are common in all the horizons.

The succession of three soils one above the other developed in eolian and colluvial materials demonstrate alternating unstable periods (with dominantly geomorphological processes) with stable periods of soil formation (with active biological processes and inactive geomorphological processes).

Vegetation community 54: *Calamagrostis epigeos* and *Carex arenaria* with *Dicranum scoparium*

Site characteristics: Elevation: 5 m; flat valley bottom; exposition: -; distance from shore line: 900 m; vegetation cover: 70% with forest remnants; soil horizons: F-H-A-E-B-C (total thickness: >12 cm).

Soil characteristics: The soil profile below vegetation 54 is characterized by organic rich horizons (F-H-A) at the top, with recent litter and root remnants, and modified organic materials in irregular aggregates and excrements, indicating a relatively high biological activity. Shell fragments and carbonate nodules of micrite, with solution characteristics at the rims, occur in the B and lower horizons. The horizon development indicates a relatively stable soil, in which other geomorphological processes are of minor importance.

5 Conclusions; future fields of micromorphological research

In dune areas micromorphological analysis could be applied in the investigation of:

- Structure and composition of dune sediments. The content of rock fragments, minerals and organic components can contribute to tracing the source area of the dune sands and to the recognition of modifications of the dune sands by weathering and soil formation after deposition. The structural differences between eolian deposits and deposits by water action are not yet clear. To obtain a better understanding of the role of carbonates in the development of certain associations of plant species, it is necessary to identify the various types of carbonate minerals. This could be done by means of chemical indicators on undisturbed impregnated slices of dune sands.

- Soil formation in dunes. The formation of biopores, the modification of fresh organic materials in amorphous humus and in excrements, the weathering of carbonates and the precipitation of secondary carbonates, and the formation of grain coatings composed of a mixture of clay and iron compounds or of organic material, should be investigated. With this knowledge we can identify the phenomena that are general and those that are specific for a certain plant association. This is necessary to understand the prerequisists for the development of a certain vegetation community at a certain spot in the dune landscape.

Cooperation between soil fauna specialists and micromorphologists is necessary in order to identify the soil organisms responsible for the various types of excrements.

- Buried soils (paleosols). Thin section analysis can be applied in distinguishing paleosols from organic rich layers (e.g. sediments) resembling soil horizons. Knowledge of paleosols can be an important contribution to the reconstruction of the paleogeomorphological history of the dunes and thus to the dynamics of dune development.

Acknowledgement

Mr. H. Snater, P.W.N., Castricum, made valuable comments on an earlier draft.

References

ASSENDORP, D. & MÜCHER, H.J. (1990): Dynamiek van de duinen weerspiegeld in een bodemprofiel met begraven bodems in het duingebied Meijendel. De Levende Natuur **91**, 2, 40–45.

BAKKER, T.W.M., KLIJN, J.A. & VAN ZADELHOF, F.F.J. (1979): Duinen en duinvalleien: een landschaps-ecologische studie van het Nederlandse duingebied. Pudoc, Wageningen, 201 pp.

BAL, L. (1973): Micromorphological analysis of soils. Lower levels in the organization of organic soil materials. Netherlands Soil Survey Institute, Wageningen. Netherlands Soil Survey Papers no. **6**, 175 pp.

BAL, L. (1982): Zoological ripening of soils. Pudoc, Wageningen, 365 pp.

BARRATT, B.C. (1962): Soil organic régime of coastal sand dunes. Nature no. **4657**, 835–837.

BOERBOOM, J.H.A. (1963): Het verband tussen bodem en vegetatie in de Wassenaarse duinen. Boor en Spade 12, 120–155.

BULLOCK, P., FEDOROFF, N., JONGERIUS, A., STOOPS, G. & TURSINA, T. (1985): Handbook for soil thin section description. Waine Research, Albrighton, Wolverhampton, England, 152 pp.

CHARTRES, C.J. & MÜCHER, H.J. (1989): The effects of fire on the surface properties and seed germination in two shallow monoliths from a rangeland soil subjected to simulated raindrop impact and water erosion. Earth Surface Processes and Landforms 14, 5, 407–417.

DUIJN, R. (1988): Micromorfologisch onderzoek van de bodemprofielen onder de vegetatietypen 54, 13K, 3K an 12K van de vegetatiekartering uit het Noordhollands duinreservaat. Laboratory of Physical Geography and Soil Science, Amsterdam; P.W.N., Bakkum. Internal Report, part 2, 44 pp.

HOLE, F.D. (1981): Effects of animals on soil. Geoderma 25, 1/2, 75–112.

JONGERIUS, A. (1957): Morfologische onderzoekingen over de bodemstructuur. Mededelingen van de Stichting voor Bodemkartering, Wageningen. Bodemkundige Studies 2, 93 pp.

JUNGERIUS, P.D. & VAN DER MEULEN, F. (1988): Erosion processes in a dune landscape along the Dutch coast. CATENA 15, 217–228.

KLIJN, J.A. (1981): Nederlandse kustduinen; geomorfologie en bodems. Pudoc, Wageningen, 188 pp.

KOOISTRA, M.J. (1978): Soil development in recent marine sediments of the intertidal zone in the Oosterschelde, The Netherlands. A soil micromorphological approach. Laboratory of Physical Geography and Soil Science, University of Amsterdam, Publ. nr. 24, 183 pp. and Netherlands Soil Survey Institute, Wageningen. Soil Survey Papers no. 14, 183 pp.

KRUYSEN, B.W.J.M. (1990): Vegetatiekaart 1:5.000 van het Noord-Hollandse Duinreservaat, 1982–1985. P.W.N., Castricum.

MÜCHER, H.J., CHARTRES, C.J., TONGWAY, D.J. & GREENE, R.S.B. (1988): Micromorphology and significance of the surface crusts of soils in rangelands near Cobar, Australia. Geoderma 42, 227–244.

POELMAN, J.N.B., BOUMA, J. & WÖSTEN, J.H.M. (1989): Significance of low-chroma mottles in drained sandy soils in The Netherlands. Soil Sci. Soc. Am. J. 53, 2, 591–592.

RUSEK, J. (1975): Collembola and Acarina as factors in soil formation. Pédobiologia 15, 299–308.

SLIKKER, F.C. & STOKVIS, J.S. (1977): Een bodemkartering van de duinen van Walcheren. Rijkswaterstaat Deltadienst, Middelburg. Studentenrapport 4–77.

SLIKKER, F.C. & STOKVIS, J.S. (1979): Een fysisch geografisch landschaps-onderzoek van de duinen van Walcheren. Rijkswaterstaat Deltadienst, Middelburg/ Rijksuniversiteit Utrecht. Studenten rapport 3-79, 93 pp.

VAN DER MEULEN, F., JUNGERIUS, P.D. & VISSER, J.H. (Eds.) (1989): Perspectives in coastal dune management. SPB Academic Publishing bv., The Hague, The Netherlands, 333 pp.

VAN DER ZIJP, M. (1989): Micromorfologisch onderzoek met betrekking tot de relatie, bodem, vegetatie and fysiografische ligging in het Noord-Hollands Duinreservaat nabij Castricum. Laboratory of Physical Geography and Soil Science, University of Amsterdam, Amsterdam/P.W.N., Bakkum. Internal report, part 2, 62 pp.

ZACHARIAE, G. (1965): Spuren tierischer Tätigkeit im Boden des Buchenwaldes. Forstwissenschaftliche Forschungen (Beihefte zum forstlichen Zentralblatt), 20, 1–68.

ZONNEVELD, I.S. (1959): Het verband tussen bodem- en vegetatiekundig onderzoek. Boor en Spade 10, 38–58.

Address of author:
H.J. Mücher
Laboratory of Physical Geography and Soil Science
University of Amsterdam
Dapperstraat 115
1093 BS Amsterdam
The Netherlands

FORTHCOMING PUBLICATION DECEMBER 1990

Miroslav Kutílek and Don R. Nielsen

SOIL HYDROLOGY

- textbook for students of soil science, agriculture, forestry, geoecology, hydrology, geomorphology or other related disciplines -

CONTENTS

1. Soil hydrology as a component of the hydrological cycle.
2. Soil porosity, methods of determination, models of soil porous system.
3. Soil water content and the measuring techniques.
4. Hydrostatics of soil water. Adsorption, capillarity, soil water potential and soil water retention curves, hysteresis.
5. Hydrodynamics of soil water. Darcy's and Darcy-Buckingham's equation, K and D functions, Richard's equation.
6. Elementary soil hydrological processes: Infiltration, redistribution, drainage, evaporation and evapotranspiration.
7. Field estimation of soil hydraulic functions.
8. Field soil heterogenity, space and time variability of soil hydraulic functions, scaling, geostatistics.
9. Transport of solutes in soil, dispersion, conservative and non-conservative transport, field solutions.
10. Models in soil hydrology: Empirical, deterministic, stochastic. Classification of soil water regimes.

ORDER FORM

☐ Please send me .... copies of: **SOIL HYDROLOGY**, CATENA paperback 1990, 250 pages, ISBN 3-923381-26-3, at the rate of DM 39.-/ US $ 24.-

Advance payment is required.

Name ....

Address ....

Date ....

Signature: ....

Please charge my credit card: ☐ MasterCard/Eurocard/Access ☐ Visa ☐ Diners ☐ American Express

Card No.: .... Expiration date: ....

Signature: ....

Please, send your orders to:

CATENA VERLAG, Brockenblick 8, D-3302 Cremlingen-Destedt, West Germany, tel.05306-1530, fax 05306-1560

USA/Canada: **CATENA VERLAG**, P.O.Box 368, Lawrence, KS 66044, USA, Tel. (913) 843-1235, fax (913) 843-1274

Japan: **EASTERN BOOK SERVICE**, 37 - 3 Hongo chome - 3, Bunkyo-ku, Tokyo 113, Japan, tel. (03) 818-0861, fax (03) 818-0864

WATER REPELLENCY IN THE DUNES WITH SPECIAL REFERENCE TO THE NETHERLANDS

L.W. **Dekker**, Wageningen
P.D. **Jungerius**, Amsterdam

Abstract

Depth, degree and spatial variability of water repellency were examined in the surface layers of the coastal dune sands of The Netherlands. The methods used include the water drop penetration time (WDPT) test. The degree of water repellency is not dependent on calcium carbonate content of the dune sand, but the character of the vegetation and the organic matter content play an important role.

The spatial variability of water repellency and, accordingly, soil wetting is extremely high, both on the surface and within the soil profile. It appears that, in contrast to what is generally believed, dune sand acts not as an isotropic medium with respect to infiltrating water. This has serious consequences for the filtering function of dune sand because transport of solutes to the groundwater reservoir along preferential flow paths is much more rapid than is assumed in hydrological models.

ISSN 0722-0723
ISBN 3-923381-23-9
©1990 by CATENA VERLAG,
D–3302 Cremlingen-Destedt, W. Germany
3-923381-23-9/90/5011851/US$ 2.00 + 0.25

1 Introduction

Water repellency of surface horizons is an extensive, often unobserved, property of soils that frequently dry out. The simplest recognition of soils with a water repellency problem is adding a drop of water to the surface of a fairly dry soil. If water upon initial contact with the soil "beads up" into a spherical shape instead of quickly being absorbed into the soil, the soil is water repellent.

Water repellent soils can be found in many parts of the world under a variety of climatic conditions (DEBANO 1981). In The Netherlands water drop penetration time (WDPT) was measured by the Soil Survey Institute on a great number of soil samples collected throughout the country. The measurements revealed that more than 75% of the agricultural topsoils are slightly to extremely water repellent and of the topsoils in nature reserves, including dunes, more than 95% exhibit strong to extreme water repellency (DEKKER 1988). Water repellency in dunes is not restricted to The Netherlands; the authors have observed the phenomenon all along the coast of Western Europe.

Numerous laboratory and field studies have been conducted to investigate the physics of water flow in water repellent soils (DEBANO 1981, HENDRICKX et al. 1988a and 1988b, LETEY et al. 1975, VAN OMMEN et al. 1988). These studies indicate that infiltration rates into water repellent soils can be considerably lower than those into wettable soils, and that wetting patterns in water repellent soils can be quite irregular and incomplete. The studies also suggest that water repellency has its greatest effect in relatively dry soils. This may be one factor involved in the observed larger-percent runoff from rainstorms as compared to later, comparable storms (LETEY et al. 1975, BRIDGE & ROSS 1983, RUTIN 1983). Thus, water repellency tends to increase runoff and erosion and to decrease available water holding capacity in affected horizons.

As a consequence of water repellency water and solutes often move in these soils through preferential flow paths, the so called "fingers" or "tongues". This phenomenon shortens solute traveltime and —in case of toxic substances— increases risk for groundwater pollution as was found by HENDRICKX and his coworkers (1988a) for a water repellent dune sand in the southwestern part of the Netherlands.

In this study we examine the depth, degree and spatial variability of water repellency in the surface layers of the coastal dune sands of The Netherlands. In addition, we want to investigate the influence of the type of vegetation and the calcium carbonate content of the dune sands on the degree of water repellency.

2 Factors affecting water repellency

Water repellent soils have been investigated in The Netherlands only recently (RUTIN 1983, DEKKER 1985 and 1988, HENDRICKX et al. 1988a). Elsewhere in the world the phenomenon of water repellency in soils has been recognised for many years and studied at different times by various workers. These studies indicate that the extent of water repellency in a soil is dependent on soil factors which affect fungal proliferation and litter decomposition, pH, texture, etc. Repellency is caused by a range of hydrophobic organic materials. In the sandy soils of southern Australia these include fungal hyphae (BOND & HARRIS 1964), humic acids (ROBERTS & CARBON 1972) and decomposing plant material or litter (McGHIE 1987). Water repellency can be found under a variety of vegetation types including forests, brushfields, grasslands, agricultural lands, and on golf greens (DEBANO 1981). Different types of plant covers produced different degrees of water repellency; in Australia the order of decreasing water repellency was phalaris, mallee, heath, and pine (DEBANO 1981). ROBERTS & CARBON (1972) have shown that extracts of various plants are capable of causing water repellency in treated sand. Because strong alkali is well documented in its ability to remove hydrophobic materials, humic acids have been regarded as the most likely candidates for the hydrophobic factors (ADHIKARI & CHAKRABARTI 1976, TSCHAPEK 1984). However, not all humic acids produce water repellency (SAVAGE et al. 1969) and the possible cleavage of covalent linkages under those conditions makes it difficult to interpret what is happening in the intact situation. MASHUM & FARMER (1985) have provided evidence to indicate that molecular orientation of organic matter determines whether or not a soil is water repellent. MASHUM et al. (1988) investigated the use of amphiphilic solvent mixtures in soxhlet extraction procedure to remove hydrophobic materials from water repellent soils. The extracted materials have been characterized by chromatographic and spectroscopic techniques and shown to contain extensive polymethylene chains including both long chain fatty acids and esters.

3 Characteristics of the area of investigation

The experimental sites are located in the coastal dune area in the western part of The Netherlands and consist of sand soils of aeolian origin. A field study to examine the occurrence and distribution of water repellent sandy dune soils was established between Den Helder and IJmuiden in the northwestern part of the country.

The younger dunes were mainly formed between 1200 and 1600 A.D and are usually 20–30 m high, consisting of an almost continuous foredune at the seaward side and of mainly parabolic dunes further landward. The foredune rises steeply from the beach to about 15–20 m above sea level.

At present nearly all dunes are covered with vegetation. Only locally, mainly near the beach, dunes without or with sparse vegetation occur. North of Bergen the dune sands generally have a low $CaCO_3$ content, mostly below 0.5%. Usually the foredune has a slightly higher lime content than the dunes farther inland, probably due to admixture of fresh shell fragments from the beach. South of Egmond the lime content of the dune sands is generally high (2–4%) and may increase locally to more than 25% through sorting, as shell fragments, being heavier and usually larger than sand grains, tend to accumulate in lag deposits (EISMA 1968).

In the dunes the variable composition of the sands has resulted in the development of very different types of vegetation, which is partly due to the amount of calcium carbonate present.

4 Methods

4.1 Soil sampling

Throughout the dune sand area between Den Helder and IJmuiden soil was sampled to a depth of 50 cm under a variety of vegetation types. At more than 500 locations soil samples were taken to a depth of 50 cm in autumn of 1988 and in spring of 1989. The samples were collected with an auger at depth 0–5, 5–10, 10–20, 20–30, 30–40 and 40–50 cm. At the laboratory the samples were dried for several days at 60°C, an estimation was made of the organic matter content and the presence of lime was checked by applying some drops of acid (10% HCl solution). On the smoothed surface of the samples the degree of water repellency was measured.

4.2 Water Drop Penetration Time (WDPT) test

Water repellency was measured with the water drop penetration time (WDPT) test used and described by several workers (e.g. HAMMOND & YUAN 1969,

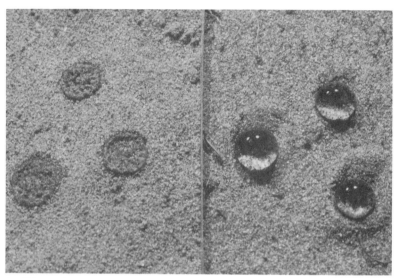

Photo 1: *Water drops penetrate immediately in wettable soil and can stay for hours on the surface of extremely water repellent soil.*

LETEY et al. 1975, RICHARDSON 1984). Three drops of distilled water from a standard medicine dropper (approximately 6 mm diameter) were placed on a smoothed air-dry surface of a soil sample and the length of time to penetrate the soil was timed (photo 1).

In general, a soil is considered water repellent (contact angle >90 degrees) if the water drop penetration time exceeds 5 sec (e.g. BOND & HARRIS 1964, DEBANO 1981, RICHARDSON 1984). The 5 sec time period was selected for convenience and does not have any specific physical meaning (RICHARDSON 1984).

After removing samples from the stove, WDPT test was deferred at least two days to obtain samples in equilibrium with ambient air humidity. We used the WDPT of the second drop (the median value) for the classification of the water repellency of the sample. We timed water drop penetration time up to 3600 sec. We distinguished the following classes: wettable, non-water repellent (<5 s); slightly (5–60 s); strongly (60–600 s); severely (600–3600 s); and extremely water repellent (>3600 s).

4.3 Soil water content measurements

Ten times during the period April 1988 through February 1989 samples for the determination of the volumetric soil water content have been taken on a parcel of dune sand with grasscover. The parcel, used for extensive agriculture, is located at Ouddorp in the southwestern part of The Netherlands, and has a rather flat surface. The samples have been taken at five depths from the soil surface to 50 cm depth with steel cores (100 cm^3) after digging trenches. At each depth hundred samples were taken in close order over a distance of 550 cm. The steel cores were emptied in plastic bags and used again. The wet soil in the plastic bags was weighed, dried for several days at 60°C, and weighed again for determination of soil water content and bulk density.

4.4 Visualizing preferential flow paths

The penetration of rain in some dune sands was also studied by using the dyestuff staining technique described by BOND (1964). Trenches were dug during the autumn and winter of 1988 after rains. The pit face was dusted with a dry mixture of 1% Rhodamine B in finely ground kaolinite until the face was uniformly covered with the white powder. Within a few minutes the wet areas developed an intense red colour whereas the dry areas remained white. The pattern was photographed on colour film to give a permanent record for comparison.

5 Results and discussion

5.1 Influence of the calcium carbonate content on the water repellency of the coastal dunes

The dune sands north of Bergen are non-calcareous throughout the profile with exception of the present foredune along the beach where calcareous sands occur (fig. 1). South of Bergen nearly all dune sands are calcareous within 5 cm depth, only locally the toplayer is decalcified up to sometimes 50 cm depth (fig. 1). As can be seen on the second map of fig. 1 most dune sands north as well as south of Bergen have a severely to extremely water repellent toplayer (0-5 cm depth). Only the toplayers of the calcareous dune sands in the foredunes are mostly wettable or slightly to strongly water repellent.

The distribution pattern at 40-50 cm depth on the third map of fig. 1 indicates that there also is no significant difference in the thickness of water repellent layers between the non-calcareous and the calcareous dune sands. In summary, there is no apparent effect of the acidity or alkalinity of the soil on water repellency as calcareous sands and acid sands are equally affected.

5.2 Influence of the type of vegetation and organic matter content

There are differences in the degree of water repellency of the dune sands due to the type of vegetation as is shown in fig. 2. All samples were taken in the dune area between Den Helder and IJmuiden. Samples of fig. 2a were taken at 104 locations in a strip of the foredunes with a sparse marram-grass (*Ammophila arenaria*) vegetation. Samples of fig. 2b were collected at 48 locations on calcareous and non-calcareous dune sands with a grey hair-grass (*Corynephorus canescens*)/moss vegetation; samples of fig. 2c are derived from 42 locations on non-calcareous sands with a heather (with *Calluna vulgaris* and *Empetrum nigrum*) vegetation and samples of fig. 2d from 36 locations with a buckthorn (*Hippophaë rhamnoides*) vegetation on calcareous sands.

The upper 20 cm of the dune sands under grey hair-grass, heather and buckthorn are mostly strongly to extremely water repellent. Locally thick layers up to 50 cm with severe to extreme water repellency occur under a heather and buckthorn vegetation. Obviously, there is no significant difference in degree and thickness of water repellent layers between the non-calcareous dune sands under heather and the calcareous sands un-

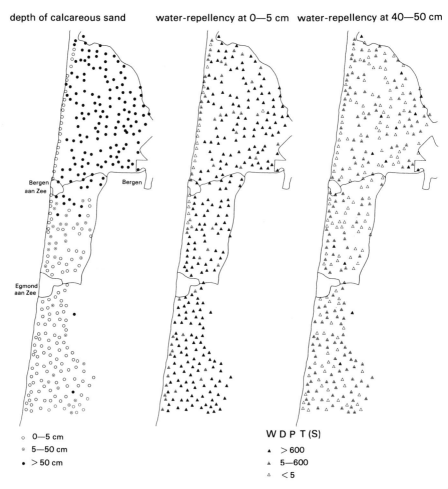

Fig. 1: *Depth of calcareous sand and degree of water repellency at 0–5 cm depth and at 40–50 cm depth in dune sands north and south of Bergen.*

der a buckthorn vegetation. The thickness of the water repellent layer under a grey hair-grass/moss vegetation is generally less than under a heather and buckthorn vegetation. More than 50 percent of the toplayers of the sands with sparse marram-grass vegetation is wettable or non-water repellent, and less than 20 percent is strongly water repellent, as is shown in fig. 2a. Remarkable is the equal distribution of the samples over the water repellency classes throughout these profiles to 50 cm depth. This is due to the low organic matter content throughout the entire profiles. The layers from 5 to 50 cm depth of more than 90% of the locations contain less than 0.5% organic matter (tab. 1).

The decrease in water repellency with depth under the other three vegetation types (fig. 2b, c, d) is closely associated with the reduced amounts of or-

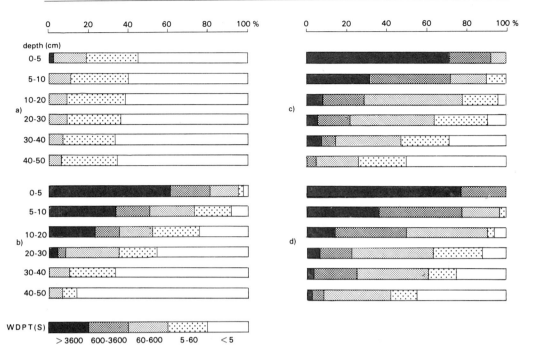

Fig. 2: *Relative frequency of the WDPT of samples at six depths in dune sands with a vegetation of a) marram-grass, b) grey hair-grass, c) heather and d) hawthorn.*

	Depth cm	Percentage of estimated organic matter					
		<0.5	0.5–1	1–2	2–3	3–4	4–8
a)	Sparse marram-grass						
	0–5	83*	15	2	—	—	—
	5–10	92*	8	—	—	—	—
	10–20	91*	9	—	—	—	—
	20–30	93*	7	—	—	—	—
	30–40	93*	7	—	—	—	—
	40–50	96*	4	—	—	—	—
b)	Various vegetation types						
	0–5	3	6	21	33*	29	8
	5–10	18	33*	35	10.5	3	0.5
	10–20	34	37*	24	4	1	—
	20–30	51*	36	13	—	—	—
	30–40	61*	30	8	1	—	—
	40–50	74*	17	9	—	—	—
c)	Grey hair-grass vegetation						
	0–5	11	6	27	37*	19	—
	5–10	34	31*	25	8	2	—
	10–20	58*	25	15	2	—	—
	20–30	83*	15	2	—	—	—
	30–40	90*	8	2	—	—	—
	40–50	98*	2	—	—	—	—

Tab. 1: *Relative frequency of the estimated organic matter content of samples collected from six depths in dune sands between Den Helder and IJmuiden a) at 104 locations in foredunes with a scarce marram-grass vegetation, b) at 403 locations with various vegetation types; c) at 48 locations with a grey hair-grass vegetation. The class with the median value of each layer is indicated with an asterisk.*

ganic matter, which would be expected if soil organic matter were the source of the hydrophobic materials. Tab. 1 gives the distribution of the estimated organic matter contents at 403 locations in the dune sands under a variety of vegetation types, excluding the sparse marram-grass along the beach coast. The average amount of organic matter in the upper toplayer is between 2 and 3% and decreases to less than 0.5% at 20–30 cm depth.

The more rapid decrease of water repellency with depth under the grey hair-grass vegetation in comparison with heather and buckthorn can also be related to the organic matter content. The organic matter content of the sand under the grey hair-grass vegetation drops to less than 0.5% within 20 cm depth (tab. 1).

5.3 Spatial variability of the degree of water repellency

The degree of water repellency, measured with the WDPT-test, can be different from place to place in the dune sands under similar circumstances as to $CaCO_3$ content and vegetation type. In the calcareous dune sands with a sparse marram-grass vegetation between Den Helder and IJmuiden, the degree of water repellency in the topsoil and in the subsoil varies from wettable to strongly water repellent, as can be seen in fig. 2. The water repellency of the dune sands with a grey hair-grass vegetation varies in the upper 30 cm of the profiles from wettable to extremely water repellent, and in the subsoil from 30 to 50 cm depth from wettable to strongly water repellent. Also the locations with a heather and buckthorn vegetation show a great variation in degree of water repellency, especially in the layers between 5 and 50 cm depth. The variation in degree of water repellency is partly due to the variation in organic matter content of the dune sand layers.

We determined the spatial variability of the WDPT over short distances at several depths on a parcel of dune sand with grasscover. The WDPT of hundred samples, collected in July 1988 at depth 45–50 cm, taken over a horizontal distance of 5.5 m, varied from less than 1 second to more than 600 seconds. At depth 25–30 cm the WDPT varied between 5 and more than 3600 seconds, and at depth 5–10 cm between one and several hours.

5.4 Implications of the water repellent dune sand

Usually, dry soil readily absorbs water because of strong attraction between the mineral particles and water. The affinity of soils for water can be reduced by hydrophobic organic materials which are either intermixed with the soil or form a coating around the mineral soil particles. Water movement can be severely limited by water repellent sandy topsoils when they are dry.

Rain which falls on the surface of a water repellent sand does not penetrate evenly. Trenches dug a few days after a good fall of rain showed that water moved downward through narrow channels, leaving the intervening soil quite dry and causing considerable variation in the moisture content of the sand (photo 2). The channels, tongues or preferential flow paths are of less resistance due to a lower water repellency or ponding on the surface in shallow surface depressions where the hydrostatic pressure aids water entry.

Water Repellency in the Dunes

Photo 2: *Preferential flow paths in dune sand visualized by using a dye stuff staining. The broken line indicates the boundary between the vertical wall and the horizontal plane of the trench.*

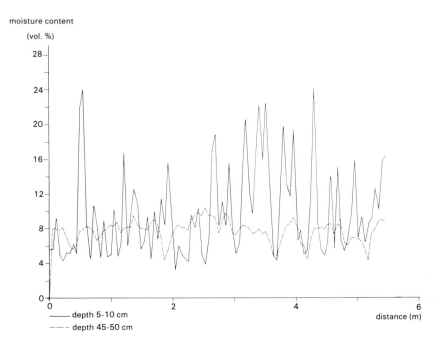

Fig. 3: *Variability of the soil moisture content in a dune sand with grasscover at two depths over a distance of 5.5 m.*

When preferential flow paths are formed, the soil no longer impedes the penetration of water, so that additional rain tends to flow down the preferential paths wich have already been wetted. Thus dry zones tend to persist and this was confirmed by periodical examination of the dune sand area throughout the winters of 1987 and 1988.

Fig. 3 shows the variation in moisture content in a dune sand with grasscover at depth 5–10 cm and at depth 45–50 cm, measured at hundred samples over a horizontal distance of 550 cm on July 12, 1988. The moisture content at depth 5–10 cm fluctuated between 4 and 24%, several times over distances of less than 5.5 cm. Clearly the highest peaks in fig. 3 with moisture contents of 15 to 24% represent the preferential flow paths. The variation in moisture contents at depth 45–50 cm is much less, and moisture contents between 4 and 10% occur.

In these soils to which one would expect one-dimensional hydrological models to apply, water and solutes appear to move through preferential flow paths. This phenomenon shortens solute traveltime and —in case of toxic substances— increases risk for groundwater pollution (HENDRICKX et al. 1988a).

References

ADHIKARI, M. & CHAKRABARTI, G. (1976): Contribution of natural and microbial humic acids to water repellency in soil. J. Indian Soc. Soil Sci. **24**, 217–219.

BOND, R.D. (1964): The influence of the microflora on the physical properties of soils. II. Field studies on water repellent sands. Australian Journal of Soil Research **2**, 123–131.

BOND, R.D. & HARRIS, J.R. (1964): The influence of the microflora on physical properties of soils. I. Effects associated with filamentous algae and fungi. Australian Journal of Soil Research **2**, 111–122.

BRIDGE, B.J. & ROSS, P.J. (1983): Water erosion in vegetated sand dunes at Cooloola, south-east Queensland. In: Jennings, J. & Hagedorn, H. (eds.), Dunes: Continental and coastal. Zeitschrift für Geomorphologie Supplementband 45, 227–244.

DEBANO, L.F. (1981): Water repellent soils: a state-of-the-art. Pacific Southwest Forest and Range Exp. Station. Gen. Tech. Rep. PSW-46. 21 pp.

DEKKER, L.W. (1985): Opname van water in moeilijk bevochtigbare zand- en veengronden. Cultuurtechnisch Tijdschrift **25**, 121–132.

DEKKER, L.W. (1988): Verspreiding, oorzaken, gevolgen en verbeterings mogelijkheden van waterafstotende gronden in Nederland. Rapport nr. 2046 Stichting voor Bodemkartering, Wageningen, 54 pp.

EISMA, D. (1968): Composition, origin and distribution of Dutch coastal sands between Hoek van Holland and the island of Vlieland. Netherlands Journal of Sea Research **4**, 123–267.

HAMMOND, L.C. & YUAN, T.L. (1969): Methods of measuring water repellency of soils. In: L.F. DeBano & J. Letey (eds.), Proc. of the symp. on water-repellent soils, 49–60. Univ. of Calif., Riverside.

HENDRICKX, J.M.H., DEKKER, L.W., VAN ZUILEN, E.J. & BOERSMA, O.H. (1988a): Water and solute movement through a water repellent sand soil with a grasscover. In: Wierenga, P.J. & Bachelet, D., Proc. Internat. Conf. and Workshop on the Validation of Flow and Transport Models for the Unsaturated Zone, Ruidoso, New Mexico, May 23–26, 1988, 131–146.

HENDRICKX, J.M.H., DEKKER, L.W., BANNINK, M.H. & VAN OMMEN, H.C. (1988b): Significance of soil survey for agrohydrological studies. Agricultural Water Management **14**, 195–208.

LETEY, J., OSBORN, J. & VALORAS, N. (1975): Soil water repellency and the use of nonionic surfactants. Calif. Water Res. Center, Contribution **154**. 85 pp.

MA'SHUM, M. & FARMER, V.C. (1985): Origin and assessment of water repellency of a sandy South Australian soil. Australian Journal of Soil Research **23**, 623–626.

MA'SHUM, M., TATE, M.E., JONES, G.P. & OADES, J.M. (1988): Extraction and characterization of water-repellent materials from Australian soils. Journal of Soil Science **39**, 99–109.

McGHIE, D.A. (1987): Non-wetting soils in western Australia. New Zealand Turf Management Journal, November, 13–16.

RICHARDSON, J.L. (1984): Field observation and measurement of water repellency for soil surveyors. Soil Survey Horizons 25, 32–36.

ROBERTS, F.J. & CARBON, B.A. (1972): Water repellence in sandy soils of South-Western Australia. II. Some chemical characteristics of the hydrophobic skins. Australian Journal of Soil Research 10, 35–42.

RUTIN, J. (1983): Erosional processes on a coastal sand dune, de Blink, Noordwijkerhout, The Netherlands. Publ. Lab. Phys. Geography and Soil Sci. Univ. Amsterdam 35, 144 pp.

SAVAGE, S.M., MARTIN, J.P. & LETEY, J. (1969): Contribution of some soil fungi to natural and heat-induced water repellency in sand. Soil Sci. Soc. Amer. Proc. 33, 405–409.

TSCHAPEK, M. (1984): Criteria for determining the hydrophylity-hydrophobicity of soils. Zeitschrift für Pflanzenernährung und Bodenkunde 147, 137–149.

VAN OMMEN, H.C., DEKKER, L.W., DIJKSMA, R., HULSHOF, J. & VAN DER MOLEN, W.H. (1988): A new technique for evaluating the presence of preferential flow paths in nonstructured soils. Soil Sci. Soc. Am. J. 52, 1192–1193.

Addresses of authors:
L.W. Dekker
The Winand Staring Centre for Integrated Land, Soil and Water Research
P.O. Box 125
6700 AC Wageningen
The Netherlands
P.D. Jungerius
Landscape and Environmental Research Group
Univ. of Amsterdam
Dapperstraat 115
1093 BS Amsterdam
The Netherlands

MODELS FOR PROCESSES IN THE SOIL

— Programs and Exercises —

by

R. Anlauf, K.Ch. Kersebaum, Liu Ya Ping, A. Nuske-Schüler, Jörg Richter. G. Springob, K.M.Syring & J. Utermann

introduced and coordinated by

Jörg Richter

ORDER FORM

☐ Please send me .... copies of: **MODELS FOR PROCESSES IN THE SOIL**, CATENA paperback 1990, 240 pages, ISBN 3-923381-24-7, at the rate of DM 38,50/ US $ 24.-

Advance payment is required.

Name ....

Address ....

Date ....

Signature: ....

Please charge my credit card: ☐ MasterCard/Eurocard/Access ☐ Visa ☐ Diners ☐ American Express

Card No.: .... Expiration date: ....

Signature: ....

Please, send your orders to:

CATENA VERLAG, Brockenblick 8, D-3302 Cremlingen-Destedt, West Germany, tel.05306-1530, fax 05306-1560

USA/Canada: **CATENA VERLAG**, P.O.Box 368, Lawrence, KS 66044, USA, Tel. (913) 843-1235, fax (913) 843-1274

Japan: **EASTERN BOOK SERVICE**, 37 - 3 Hongo chome - 3, Bunkyo-ku, Tokyo 113, Japan, tel. (03) 818-0861, fax (03) 818-0864

WATER EROSION IN THE DUNES

P.D. **Jungerius**, Amsterdam
L.W. **Dekker**, Wageningen

Summary

Dunes are subject to water erosion just like other sloping relief feature in the world. The same processes are active: raindrop splash and slope wash. Of particular importance is splash drift which occurs at times of rain when splashed sand grains are taken up and transported by strong winds. Erosion by overland flow (slope wash) is effective in the 'grey' dunes, particularly on sparsely vegetated south-exposed slopes. The cause of overland flow is the impeded infiltration shown by dune soils when dry (water repellency). Slope wash occurs as unconcentrated or as concentrated wash. In the latter case rills and alluvial fans are formed. Most of the sand comes down in one or two rain storms in summer when the water repellency of the sand is particularly high. Sediment yields higher than 2.5 kg per m^2 have been recorded. In contrast to wind erosion, water erosion leads to gradual levelling of dune topography.

1 Introduction

The landscape of the coastal dunes is commonly regarded as belonging to the domain of the wind. This is justified for dunes which are presently being formed, but not for the dunes which have been stabilized and are now largely fossilized as aeolian features. Where slopes are steeper than about 6°, dunes are subject to water erosion just like other sloping terrain forms in the world. The same processes are active and the same factors as used in the Universal Soil Loss Equation apply: characteristics of the rainfall, properties of the soil, nature of the vegetation cover, and length and steepness of slope (WISCHMEIER & SMITH 1965).

Reports of water erosion in the dunes are scarce. BIGARELLA (1975) and BRIDGE & ROSS (1983, see also THOMPSON 1983) report on water erosion in Brasilia and Australia, respectively, but give no earlier references. In Europe, most work appears to have been done in The Netherlands. BERGMEIJER (1981) was probably the first to describe the water erosion processes in any detail, followed by RUTIN who carried out a two-year monitoring programme (1983).

The traces of water erosion in the dunes are much less persistent than those of wind erosion which makes it difficult to assess the extent of the process. Some of the features can be observed only during or shortly after the rain. Sand grains adhering to plants and piling up behind obstacles on slopes testify to the importance of raindrop splash, whereas

ISSN 0722-0723
ISBN 3-923381-23-9
©1990 by CATENA VERLAG,
D-3302 Cremlingen-Destedt, W. Germany
3-923381-23-9/90/5011851/US$ 2.00 + 0.25

the presence of ephemeral rills and fans points to the action of running water. In fact, erosion studies in the dunes along the coast of The Netherlands have indicated that the affected surface is large and sediment yield is high: up to 3 kg/m^2 in one storm which would be an alarming figure in agricultural land.

Erosion maps show that not all dune sand is sensitive to erosion by surface wash. The clean, yellow or white (Munsell Soil Color Charts code 8/2 for value/chroma) sand of the original aeolian deposition is as sensitive as any loose material to splash, but it is very permeable and little affected by slope wash, whereas humic, grey sand when dry is water repellent and sensitive to erosion by running water (see DEKKER & JUNGERIUS this volume). Slope wash is accordingly most important in the socalled 'grey dunes' which have been vegetated for some time (JUNGERIUS this volume, fig. 1).

Although water erosion processes in the dunes are essentially similar to their counterparts on slopes elsewhere, some of the features produced have an unusual character due to the specific conditions prevailing in the dunes. These features are described here. The geomorphological consequences are evident from the changes in the original aeolian relief.

2 Water erosion features

2.1 Raindrop splash

The effectiveness of raindrop impact, or splash, on sloping terrain depends very much on the vegetation cover. Wherever the ground is bare, the characteristic imprints are found after rain. As the vegetation cover in the dunes changes throughout the year so does the amount of splashed sand per unit rain. RUTIN (1983) found that the amount of sand moving into his splash traps was largely accounted for by the proportion of bare sand exposed in front of the traps, and the amount of rainfall during the weekly measuring period. The same two terms appear in a splash model developed by BRIDGE & ROSS (1983) for Australian coastal dunes with low shrubby woodland and open grassy forest and a mean annual rainfall ranging between 1200 and 1700 mm (THOMPSON 1983). A third term in their model is the proportion of monthly rain falling at intensities above 25 mm/h. These intensities are not a monthly feature of rainfall along the European coast. Surprisingly these authors detected no effect of differences in slope angle on sand movement.

Further complications arise from the differences in resistance offered by the nature of the soil surface which depends on the presence of dead and living organic matter on and in-between the sand grains.

Three forms of displacement increase the effectiveness of raindrop splash in the dunes. The first of these is dependent on the force of the wind at the time of rain and has been called splash drift (RUTIN 1983). This process occurs when the upward movement of splashed sand already skewed by the slanting rain, is followed by wind drag. The sand particles leave their short splash path and are further transported in saltation in the direction of the wind. Splash drift is particularly active during the period of gusts which often marks the beginning of a rain storm but continues when the sand is wet. Under these conditions the rate of transport is lower because wet grains adhere to each other and are more dififcult to transport than dry grains.

Water Erosion

Photo 1: *The effect of splash drift.*

Splash drift particularly affects the clean sand of the original aeolian deposit, even in the foredunes (photo 1). It is responsible for the fact that appreciable wind erosion has been recorded on rainy days (JUNGERIUS et al. 1981). Where wind blows upslope, it counteracts the downslope gravity pull so that splashed sand particles may be transported uphill instead of downhill (ELLISON 1944). This poses a difficult problem to field measurement of the splash process in dunes.

The two other forms of displacement can be observed when the first raindrops of a storm hit water repellent sand on a slope. Some of the drops bounce up and take a small, saucer-shaped lump of sand with them in their splash path. Other drops 'ball up' and roll downslope covered in sand (photo 2).

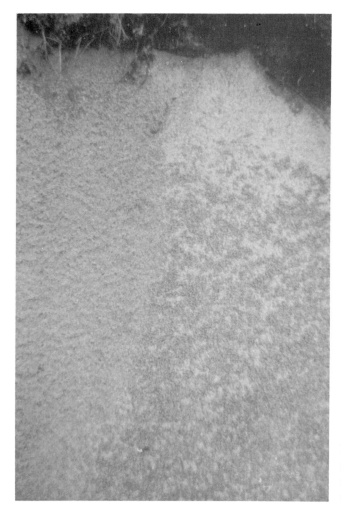

Photo 2: *Rain splash on water repellent (right hand side) and non-water repellent sand.*

2.2 Slope wash

Overland flow under water repellent conditions is essentially of the Hortonian type because it is a function of rainfall intensity, infiltration capacity and slope position. Once the sand is moist throughout its hydraulic conductivity is so high that no overland flow will develop.

Erosion of dune sand by overland flow may occur as unconcentrated or as concentrated wash. In the latter case rills and alluvial fans are formed. Most rills left by slope wash disintegrate within a few days after a shower. The alluvial toeslopes are more stable. They have a characteristic slope angle of 6°. First the sand is deposited, then lower down the organic material that has been washed out (photo 3).

HULSHOFF et al. (1986) followed the development of three 'permanent' rills in the dunes near Bergen (for location see Appendix) with erosion pins during one year. The length of the rills var-

Water Erosion

Photo 3: *Alluvial fan.*

Photo 4: *Mudflow tongues.*

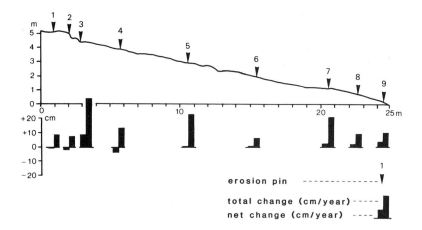

Fig. 1: *Dynamics of a rill in the dunes near Bergen (Nov. 1982–Nov. 1983).*

ied from 17 m on a 16° slope to 25 m on a 11° slope. At no place was only erosion recorded: erosion and sedimentation alternated at each erosion pin site. Fig. 1 shows the total and net amount of change of one year along the bottom of one of the rills. Maximum net incision recorded here after one year was more than 3.3 cm. Although they are erosional features, it appears that the rills have primarily a transport function.

Unconcentrated slope wash is particularly effective on the often sparsely vegetated south-oriented slopes where grey sand is exposed at the surface. This process accounted for 90% of the total sediment yield (including erosion by wind) of a dune slope near Noordwijkerhout (RUTIN 1983). More than half of the bulk of the sand came down in one rain storm of 43 mm which occurred after a dry spell in summer when the water repellency of the sand was particularly high. The importance of the vegetation is evident from the equations given by WITTER et al. (1990). Water erosion is much lower on a moss-covered surface although this plot produced much more runoff. This is presumably the result of the higher organic matter content which works in two opposite directions: it provides the soil with a higher degree of water repellency but also with a better structure.

2.3 Mudflow tongues

The water repellency of the surface 2 or 3 mm of bare, grey sand is less than lower down in the profile (JUNGERIUS & DE JONG 1989) which is presumably due to the presence of hydrophilic algae (PLUIS & DE WINDER 1989) or to the stripping of the water repellent coating from the surface of the sand grains by deposition or attrition. Whatever the cause, it has great consequences for slope wash. Water infiltrates the thin surface layer relatively easy before it meets the strongly repellent sand below. Any volume of rainwater in excess of the storage capacity of the surface layer will be discharged across the surface. The saturated upper layer is easily entrained downslope as a waterlogged slurry much

Main landscape		map symbol	relief	slope angle °	asp.	height m	vegetation
I	Foredunes	1/4	steep slopes	>15		10–20	open tall grassland, (dwarf-) shrubland
		5/6	plateaus	0–15		10–20	open tall grassland, (dwarf-) shrubland
II	Parabolic dunes	7	steep slopes	>15	S	5–25	open tall grassland, low pioneer vegetation
		8/10	steep slopes	>15	N	5–25	dense short grassland with lichen, shrubland
		11/12	small valleys	>10		2–5	dense short grass/mossland, reed land, shrubland
V	Dune valleys	5/27	undulating	0–10		2–10	grass/mossland, shrubland
		8/30	former farmland	0–5		2–10	grassland, tall shrubland decid. forest
III	High inner dunes	13	steep slopes	>15	S	10–25	open tall grassland, low pioneer vegetation
		16	steep slopes	>15	N	10–25	tall shrubland, scattered trees
		18	gentle slopes	0–10		10–35	short vegetation with moss and lichen
		19	small valleys			-10	forest, shrubland

Tab. 1: *The landscape units of fig. 2, after VAN DER MEULEN & VAN HUIS (1985).*

resembling a mudflow (RUTIN 1983). It can be calculated that the sand mobilized in this way on the dune slope amounts to around 1.5 kg per mm depth and per m² surface area. This compares well with the sediment yield of 2.5 and 3 kg/m² recorded by RUTIN (1983) and WITTER et al. (1990), respectively, during extreme storm events.

The tongues of sand moving downslope lose their water by percolation into the surface across which they pass and this will eventually bring them to a halt. The next tongue will override the preceding ones and reach farther downslope. The result is a fan-shaped set of tongues with black outlines because the organic matter is washed off during transport and is deposited below the sand because its weight is less (photo 4). A similar process was described by THOMPSON (1983) for Australia.

The sand which eventually reaches the base of the slope is laid down as colluvium. The entrained seeds germinate easily here because the soil material has a relatively high organic matter content and favourable moisture conditions. Colluvial layers of up to nearly 8 cm depth can be formed in this way within the year (HULSHOFF et al. 1986). Variations in the ratio of the speed of colluviation to the speed of soil formation with time are expressed as buried A horizons in the soil profile (JUNGERIUS this volume).

3 Slope Development

The extent of the dune surface affected by the erosion processes can best be assessed by mapping. Such maps have been made of several dune terrains along the

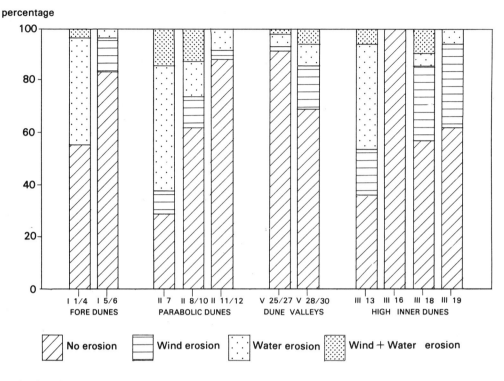

Fig. 2: *The proportion of various landscape units affected by erosion processes, in the dunes near The Hague. Relevant characteristics of the landscape units are given in tab. 1*

coast of The Netherlands (JUNGERIUS & VAN DER MEULEN 1988). To examine the type of geomorphological information they can provide, one of these maps has been summarized as shown in fig. 2. The survey was carried out in a strip of .5 km width and about 3 km length across the dunes of Meijendel (VAN DER KRAAIJ 1985). A landscape-ecological map of this terrain was made by VAN DER MEULEN & VAN HUIS (1985). From the sea shore in the west to the polders in the east, four main landscapes are recognized in the strip included in the survey: I foredunes, II parabolic dunes, V extensive dune valleys, and III high inner dunes. The erodibility of each sub-unit of the main landscapes for each of the two kinds of erosion was established in a grid of 25×25 m. VAN DIJKE & ASSENDORP (1987) processed the data for inclusion in a Geographical Information System. Comparing tab. 1 and fig. 2 shows that the two important factors are slope angle and vegetation, in that order. Water erosion affects land forms steeper than 15°, particularly south-exposed slopes with open vegetation. Type and age of the landscape nor distance to the sea appear to be important.

The water erosion processes will eventually give the dune landscape a characteristic assemblage of slopes. All slopes steeper than 15° are worn down with a speed that depends on the density of the vegetation cover. Soils on these slopes are generally poorly developed. The slope angle of the alluvial footslopes are characteristically less than 6°. The relief is gradually smoothed because higher parts are lowered and depressions are filled in. Where the vegetation cover is complete, water erosion will act as inconspicuously as in most other natural areas with a dense vegetation cover, according to the same geologic norm which has created so much of the subaerial relief of the Earth.

References

BERGMEIJER, M. (1981): Processen in kustduingebied 'De Blink', Gemeente Noordwijk. Intern rapport Fysisch Geografisch en Bodemkundig Laboratorium, Universiteit van Amsterdam.

BIGARELLA, J.J. (1975): Structures developed by dissipation of dune and beach ridge deposits. CATENA **5**, 107–152.

BRIDGE, B.J. & ROSS, P.J. (1983): Water erosion in vegetated sand dunes at Cooloola, southeast Queensland. Z. Geomorph. N.F. Suppl.-Bd. **45**, 227–244.

ELLISON, W.D. (1944): Studies of raindrop erosion. Agric. Engin. **5**, 25, 53–55.

HULSHOFF, R.M., MEER, E.H. VAN DER, SCHADE, E. & VOS, C.H. (1986): De relatie tussen erosieprocessen en vegetatie in de duinen ten zuiden van Bergen aan Zee. Intern rapport Hugo de Vries Laboratorium 201, Universiteit van Amsterdam.

JUNGERIUS, P.D., VERHEGGEN, A.J.T. & WIGGERS, A.J. (1981): The development of blowouts in 'De Blink', a coastal dune area near Noordwijkerhout, The Netherlands. Earth Surface Processes and Landforms **6**, 375–396.

JUNGERIUS, P.D. & MEULEN, F. VAN DER (1988): Erosion processes in a dune landscape along the Dutch coast. CATENA **15**, 3/4, 217–228.

JUNGERIUS, P.D. & JONG, H.H. DE (1989): Variability of waterrepellence in the dunes along the Dutch coast. CATENA **16**, 4/5, 491–497.

PLUIS, J.L.A. & WINDER, B. DE (1989): Spatial patterns in algae colonization of blowouts. CATENA **16**, 499–506.

RUTIN, J. (1983): Erosional processes on a coastal sand dune, De Blink, Noordwijkerhout. Publicaties van het Fysisch Geografisch en Bodemkundig Lab. **35**, 144 p.

THOMPSON, C.H. (1983): Development and weathering of large parabolic dune systems along the subtropical coast of eastern Australia. Z. Geomorph. N.F., Suppl.-Bd. **45**, 205–225.

VAN DER KRAAIJ, L.V. (1986): Kwetsbaarheidskartering Meijendel. Rapport Fys. Geogr. en Bodemk. Lab., Univ. van Amsterdam, t.b.v. Duinwaterleiding van 's-Gravenhage.

VAN DER MEULEN, F. & HUIS, J.C. VAN (1985): Dune landscape map of Meijendel. Dune Water Works, The Hague.

VAN DIJKE, J.J. & ASSENDORP, D. (1987): De relatie tussen wind- en watererosie, morfologie en konijnengraafactiviteit in het duingebied Meijendel, Den Haag, onderzocht met behulp van een geografisch informatiesysteem. Rapport Fysisch Geografisch en Bodemkundig Laboratorium, Universiteit van Amsterdam, 10 p.

WISCHMEIER, W.H. & SMITH, D.D. (1965): Predicting rainfall-erosion losses from cropland east of the Rocky Mountains. U.S. Dept. Agriculture Handbook **282**, Washington D.C.

WITTER, J.V., JUNGERIUS, P.D. & HARKEL, M.J. TEN (1990): Modelling water erosion and the impact of water repellency. Submitted to CATENA.

Addresses of authors:
P.D. Jungerius
Landscape and Environmental Research Group
Univ. of Amsterdam
Dapperstraat 115
1093 BS Amsterdam
The Netherlands
L.W. Dekker
The Winand Staring Centre for Integrated Land, Soil and Water Research
P.O. Box 125
6700 AC Wageningen
The Netherlands

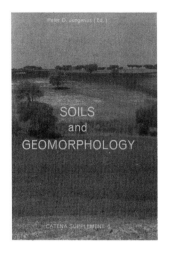

Peter D. Jungerius (Ed.):

Soils and Geomorphology

CATENA SUPPLEMENT 6 (1985)

Price DM 120,— / US $ 70.—

ISSN 0722-0723 / ISBN 3-923381-05-0

It was 12 years ago that CATENA's first issue was published with its ambitious subtitle "Interdisciplinary Journal of Geomorphology – Hydrology – Pedology". Out of the nearly one hundred papers that have been published in the regular issues since then, one-third have been concerned with subjects of a combined geomorphological and pedological nature. Last year it was decided to devote SUPPLEMENT 6 to the integration of these two disciplines. Apart from assembling a number of papers which are representative of the integrated approach, I have taken the opportunity to evaluate the character of the integration in an introductory paper. I have not attempted to cover the whole bibliography on the subject: an on-line consultation of the Georef files carried out on 29th October, 1984, produced 3627 titles under the combined keywords 'geomorphology' and 'soils'. Rather, I have made use of the ample material published in CATENA to emphasize certain points.

In spite of the fact that land forms as well as soils are largely formed by the same environmental factors, geomorphology and pedology have different roots and have developed along different lines. Papers which truly emanate the two lines of thinking are therefore relatively rare. This is regrettable because grafting the methodology of the one discipline onto research topics of the other often adds a new dimension to the framework in which the research is carried out. It is the aim of this SUPPLEMENT to stimulate the cross-fertilization of the two disciplines.

The papers are grouped into 5 categories: 1) the response of soil to erosion processes, 2) soils and slope development, 3) soils and land forms, 4) the age of soils and land forms, and 5) weathering (including karst).

P.D. Jungerius

P.D. JUNGERIUS
  SOILS AND GEOMORPHOLOGY

The response of soil to erosion processes
C.H. QUANSAH
  THE EFFECT OF SOIL TYPE, SLOPE, FLOWRATE AND THEIR INTERACTIONS ON DETACHMENT BY OVERLAND FLOW WITH AND WITHOUT RAIN
D.L. JOHNSON
  SOIL THICKNESS PROCESSES

Soils and slope development
M. WIEDER, A. YAIR & A. ARZI
  CATENARY SOIL RELATIONSHIPS ON ARID HILLSLOPES
D.C. MARRON
  COLLUVIUM IN BEDROCK HOLLOWS ON STEEP SLOPES, REDWOOD CREEK DRAINAGE BASIN, NORTHWESTERN CALIFORNIA

Soil and landforms
D.J. BRIGGS & E.K. SHISHIRA
  SOIL VARIABILITY IN GEOMORPHOLOGICALLY DEFINED SURVEY UNITS IN THE ALBUDEITE AREA OF MURCIA PROVINCE, SPAIN

C.B. CRAMPTON
  COMPACTED SOIL HORIZONS IN WESTERN CANADA

The age of soils and landforms
D.C. VAN DIJK
  SOIL GEOMORPHIC HISTORY OF THE TARA CLAY PLAINS S.E. QUEENSLAND
H. WIECHMANN & H. ZEPP
  ZUR MORPHOGENETISCHEN BEDEUTUNG DER GRAULEHME IN DER NORDEIFEL
M.J. GUCCIONE
  QUANTITATIVE ESTIMATES OF CLAY–MINERAL ALTERATION IN A SOIL CHRONOSEQUENCE IN MISSOURI, U.S.A.

Weathering (including Karst)
A.W. MANN & C.D. OLLIER
  CHEMICAL DIFFUSION AND FERRICRETE FORMATION
M. GAIFFE & S. BRUCKERT
  ANALYSE DES TRANSPORTS DE MATIERES ET DES PROCESSUS PEDOGENETIQUES IMPLIQUES DANS LES CHAINES DE SOLS DU KARST JURASSIEN

NATURAL STABILIZATION

J.L.A. **Pluis** & B. **De Winder**,
Amsterdam

Summary

Vegetation on erosional sites has a stabilization strategy which differs from that of vegetation on accumulation sites. The initial phase of stabilization of erosional sites by plants is brought about by algae, annuals and perennials. On accumulation sites plants dominate which are able to survive sand coverage.

This paper deals mainly with the mechanisms of blowout stabilization by algae. Two kinds of stabilization are distinguished: initial stabilization in which algae colonize large parts of the blowout within a short time and secondary stabilization which reflects the balance between algal colonization and sand dynamics. The first case concerns mainly small blowouts many of which become stabilized by algal crusts almost completely dominated by the green algae. Green algal crusts in large blowouts are generally found on the blowout flank facing north and on parts of the centre. Other orientations and grey soils outside the blowout tend to have relatively more cyanobacteria. Such a distribution pattern of species is partly the result of local high sand dynamics which inhibit algal crusts advancing in the development sequence. In addition, the role of soil water content is discussed. When algal crusts are able to develop the erosion rate is clearly lowered.

1 Introduction

In order to understand natural stabilization in dune sand areas, a distinction should be made between accumulation sites and erosional sites. The vegetation on accumulation sites is adapted to a range of levels of sand deposition. It acts as a "semi pervious fence": accumulating sand in and behind vegetation. On the other hand, on erosional sites the mechanisms of stabilizing sand by vegetation consist mostly of holding sand with hyphae and roots to prevent it from being blown away. It should be realized that erosion sites and accumulation sites are relative terms, because nowhere in the dunes is the wind only taking away or accumulating sand all the time. On erosion sites, the fluctuations result in net lowering, on accumulation sites in net increase in surface level.

In the first part of this paper, the stabilization of accumulation and erosional sites will be discussed in general terms. In the second part, the impact of colonization by phototrophic micro-organisms on the stabilization of blowouts will be demonstrated in a case study.

ISSN 0722-0723
ISBN 3-923381-23-9
©1990 by CATENA VERLAG,
D–3302 Cremlingen-Destedt, W. Germany
3-923381-23-9/90/5011851/US$ 2.00 + 0.25

1.1 Stabilization of accumulation sites (embryo dunes and secondary dunes)

Embryo dunes are mainly formed on the backshore. Sand accumulation usually begins behind some obstacle or element of roughness on the beach. On the coast of the North Sea and the Atlantic coast of Europe the first plant species which act as sand stabilizers on beaches are usually *Agropyrum junciforum* and *Ammophila arenaria* (ZENKOVICH 1967). The first geomorphological accumulation forms on the shore are classified by VAN DIEREN (1934) as "dune embryonales fundati" built up on the shore. Thanks to *Elymus arenarius* the dunes join to form a coastal ridge (dunes anticus) overgrown with *Elymeto-Ammophiletum typicum* on the windward and *E.-A. festucetosum* on the leeward side (WESTHOFF 1947).

Stabilization of embryo dunes might also be induced by microbial aggregation of sand. FOSTER (1979) noted aggregation with the appearance of grass, fungi and other micro-organisms before colonization with *A. junceiforme*. The increased aggregation was effected by filaments entangling sand grains particularly by mycorrhizal fungi.

Secondary dunes are accumulation sites of sand transported in the dune area. The dunes grow upward via vertical accretion with dune grass acting as a fence, trapping the sand. The vegetation anchors and stabilizes the dunes preventing dune migration. The best known stabilizer is marram (*Ammophila arenaria*). These plants are characterized by a long, elaborate root system, which reaches down to the freshwater table and have additional underground rhizomes that grow parallel to the upper dune surface. Maximum growth and branching of the root system is achieved when the marram grass is covered with abundant sand (CHAPMAN 1964). VAN DER PUTTEN et al. (1988) concluded from greenhouse experiments that fresh sand provides *A. arenaria* with a way to escape from harmful biotic soil factors, i.e. micro-organisms.

By studying the "catching capacity" on secondary accumulation sites NOEST (1987) found a relationship between the dynamics of the surface, in terms of surface level change, and plant species. This relationship was not manifestly causal. Low growing plants and slowly growing plants are found in places with relatively low sand dynamics. In other words, the range of fluctuations of the surface (by alternating phases of erosion and sedimentation) was matched by the growth rate and height of the plants.

The capacity to produce rhizoids in overlying sand and the rapidity of upward growth are important for the ability of moss to survive burial (BIRSE et al. 1957). Pioneer species like *Tortula ruraliformis* are of the acrocarpous type with an upright growth habitat capable of growing through small amounts of accreting sand. In more stable areas pleurocarpous mosses with spreading growth habit like *Hypnum cupressiforme* are found.

1.2 Stabilization of erosional sites

Deflation areas are located on the shore, in blowouts and in secondary dune valleys. The secondary dune valley originates from a blowout and is generally eroded down to the ground water level. High water content of the soil prevents the sand from deflation and stimulates growth of vegetation (photo 1).

Natural Stabilization

Photo 1: *A pioneer vegetation has colonized the basin of a secondary dune valley.*

The most important stabilization strategy for vegetation at erosional sites results from the restriction of grain movement. On parts of the shore lacking biogenic dune formation the sand might be stabilized by the formation of salt crusts or microbial mats. BOUGHEY (1957), MORTON (1957) and PYE (1980) considered that salt crusting on tropical beaches can be an important factor contributing to the limitation of dune growth. On temperate beaches salt crusts can easily be broken by impacting grains coming from areas without crusts (SVASEK & TERWINDT 1974). Laminated microbial sediment ecosystems, often referred to as microbial mats, develop best on mudflats and in the intertidal zones of beaches (HERBERT 1985, STAL et al. 1985) but are observed also on sandy beaches (DE WIT et al. 1988). Microbial mats drastically increase the stability of the sandy sediments by increasing their resistance to water movement and wind erosion (GRANT & GUST 1987).

In coastal dunes the initial stabilization is brought about by phototrophic micro-organisms forming non-laminated algal crusts (VAN DEN ANCKER et al. 1985). The algal crust lowers the energy impact of wind by increasing the surface resistanc. This in turn enables the colonization of annuals and perennials which lowers wind energy further. The magnitude and frequency of deflation and accumulation on erosional sites is commonly such that *A. arenaria* is not able to grow (VAN DER PUTTEN et al. 1988). In some cases *A. arenaria* and *Carex arenaria* extend inwards from the perimeter towards the centre of a blowout using root stocks. On the other hand colonization by *Sedum acre* is less affected by sand transport. The colonization rate of this plant species depends almost entirely on overland flow due to rainwater. Pieces of *S. acre* are easily loosened from

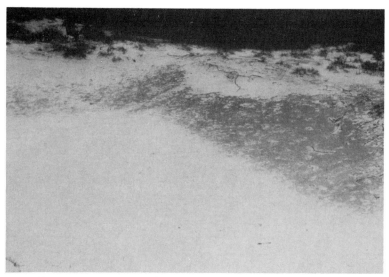
Photo 2: *Mature algal crust covering part of the blowout.*

the soil by trampling pressure.

The various mechanisms of propagation are responsible for different spatial settling patterns. Colonization of *A. arenaria* and *C. arenaria* inside a blowout commonly occurs by extending respectively from the accumulation zone and from other parts of the blowout surroundings. In respect of the prevailing SW wind, algal crusts are commonly located on the upwind inner flank of the blowouts which lack a clear boundary (PLUIS & DE WINDER 1989), whereas *S. acre* is found more in the lowest part of the blowouts, commonly being the centre.

Annuals and perennials are able to withstand erosion to a greater depth than algal cursts thanks to the fac that higher plants are anchored to a greater depth by their roots. Remnants of the roots can be observed at the blowout surface after being exposed by wind erosion. On the other hand, algae no longer contribute to the stabilization of the sand when the thin crust is eroded. However algal colonization occurs more frequently and commonly covers larger areas in blowouts than other plants.

2 Case study: Natural stabilization of blowouts by algae

2.1 Introduction

Micro-organisms play an important role in the early stabilization of drifting sand. The vegetation development on eroded surfaces often begins with phototrophic micro-organisms (GRAEBNER 1910, REYNAUD & ROGER 1981, VAN DEN ANCKER et al. 1985). In the coastal dunes near The Hague algal cursts in blowouts are build up by cyanobacteria and green algae (PLUIS & DE WINDER 1989). Cyanobacteria are considered to be the first colonizers, to be succeeded by the green algal *Klebsormidium flaccidum* which becomes dominant, and a thick crust may be found (photo 2). Succession is partly explained from the results of experiments using two cyanobacterial species and the green alga *K. flaccidum*. It was demonstrated that cyanobacteria seem to be better adapted to rapidly changing water availability wheras *K. flaccidum* seems to be related to conditions of improved water retention which is brought about by the pioneering cyanobacteria (DE WINDER et al. 1989).

This case study will discuss the contribution of the different species of phototrophic micro-organisms to the process of extending algal crusts and to the stabilization of blowouts.

2.2 Materials and methods

Data collecting from blowouts started in April 1986 in Meijendel, a coastal dune area located near The Hague. The blowouts are situated about 1350 meter from the coastline, in an area of ca. 4 ha where, in the interest of nature conservation, no stabilization measures have been taken since 1978.

This paper concerns three large blowouts (L1, L2 and L3) and three smaller blowouts (S1, S2 and S3). In 1986 the blowouts L1, L2 and L3 (downwind long-axis 85, 70 and 90° respectively) were 23, 19.3 and 18.6 metres in length. Blowouts S1, S2 and S3 (downwind long-axis 59, 102 and 106° respectively) had maximum lengths of about 2.9, 8 and 5.6 metres.

The surface of the blowouts was divided into a grid of 2×2 m squares. In the centre of each square an erosion pin was placed which recorded changes in the elevation of the surface. Small soil samples from the surface were examined for phototrophic micro-organisms using phase-contrast microscopy. Cyanobacteria were classified using the generic assignments of RIPPKA et al. (1979). Because *Microcoleus* is easily determined in field material this species is mentioned separately and not considered to belong to the LPP-group (Lyngbya, Plectonema and Phormidium). Biomass was subsequently quantified using total chlorophyll extracted with methanol (specific adsorption coefficient of 71.9 g·l^{-1}) and after Febr. 1988 with dimethyl formamide (specific absorption coefficient of 72.1 g·l^{-1}). Changes of the surface of the small blowout were recorded by photography at the time of sampling. Data were collected with a frequency of about 10 times a year, from April 1986 to March 1989.

2.3 Results

2.3.1 Small blowouts

Fig. 1 illustrates the development of the algal mat during the stabilization of blowout S1. This blowout was formed during the spring of 1987. The pins reflect sand deposition in the summer which was probably caused by slope wash. The cyanobacteria were the initial colonizers, mostly members of the genera *Microcoleus*, *Oscillatora* and *Tychonema*. Under the subsequent stable conditions an algal mat dominated by the green alga *K. flaccidum* developed above the sand surface. Gradually sand grains and small fragments of plant debris in various stages of decomposition were deposited on and in this mat and cemented into a firm crusts up to 3 mm thick. The biomass content reached values of more than 30μg Chl/cm^2. It is noted in fig. 1 that the sand movement in

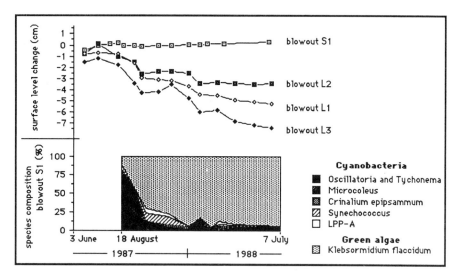

Fig. 1: *The upper part of the figure shows the mean height of the surface level in blowout S1, L1, L2 and L3 for 4, 28, 40 and 59 pins respectively. The lower part of the figure shows the species composition of the algal mat in blowout S1 illustrating the succession after initial algal colonization.*

blowout S1 almost totally stopped while other blowouts still revealed periods with surface lowering. Only at extreme high windspeed there was some erosion in blowout S1.

Algal colonization is concentrated just below the top surface of the sandy soil. This makes it difficult to observe initial growth. Increase in biomass or change in algal population therefore is not always confirmed by an increase in the area covered by algae visible with the naked eye. The algal layer becomes most clear after the wind has blown away the top sand layer. During the three years of the study period colonization of algae could be observed most clearly at the end of the summer of 1987, when, in all blowouts, the increase of the area covered by algae was most pronounced. The spatial extent of algal crusts in the blowouts ranged from approximately half the area bare sand being covered by new algal crusts (blowout L3, S2 and S3), to more than half in blowout L1 and L2, and total cover of blowout S1. Apart from blowout S1 the crusts lai scattered throughout the blowouts. Coherent parts of new mats containing cyanobacteria did not cover more than about 4 m². The bare patches revealed more surface lowering than the algal mats so that the surface level of the blowout was very irregular. In blowout S1 the initial crust developed further during the subsequent months. Colonization in this manner is most manifest in small (new) blowouts of which many are stabilized within one year.

The contribution of algae to the natural stabilization of blowouts is most clear in small blowouts. In these, green algal crusts gradually become dominant and subsequent erosion is inhibited for a long period of time. At areas which are cov-

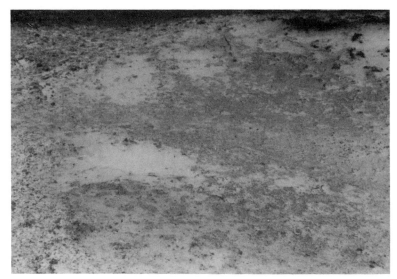

Photo 3: *At the upwind side of blowout S2 wind erosion patches were formed in the algal crust during strong wind and enlarged towards the centre of the blowout in the spring of 1987. During the summer these patches were stabilized again.*

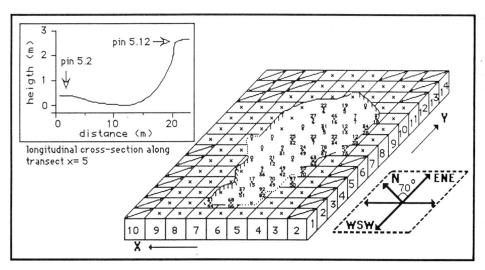

Fig. 2: *Schematic view of blowout L2 based on field observations in April 1986. The values at the pins are based on a period of three years. The upper values indicate the period of the time during which algae are present and the lower values indicate the average amount of filaments belonging to green algae compared to the total algae present. Both values are given in percentages.*

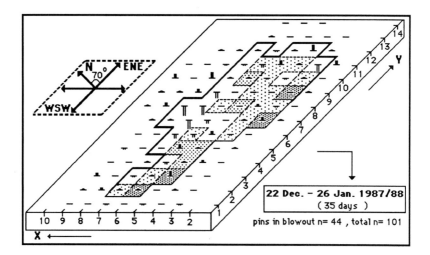

Fig. 3: *Spatial distribution of the three algal crust types in blowout L2 and the deflation rate. High erosion rates on the steep inner flank of the blowout inhibits algae colonization there. The algal crusts at the rim of the blowout has reached the mature green alga phase while the crusts in the centre are mainly dominated by cyanobacteria. For legend see fig. 5.*

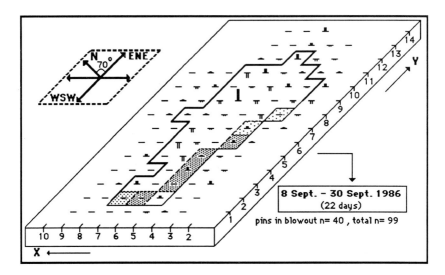

Fig. 4: *Algal crusts in blowout L2 at the end of September 1986. During the summer of 1986 the algal crusts at the rim of blowout were able to resist destruction and could reach the mature stage during the periods thereafter. For legend see fig. 5.*

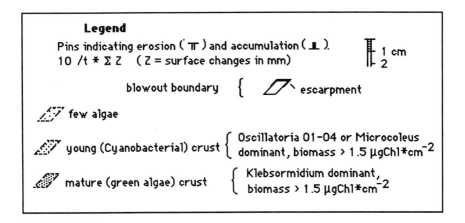

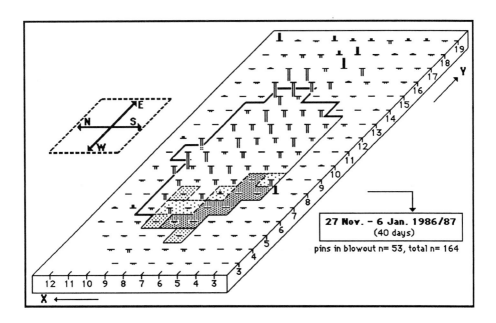

Fig. 5: *Algae in blowout L3 during part of the time when algae were able to stabilize the west side of this blowout.*

ered by algal crusts the shear threshold value is increased such that sand transport is lowered. This prevents destruction of new algal crusts. The influence of a surviving algal crust on formation of new crusts is illustrated for blowout S2. With the first field observation early in 1987, most of the surface in the blowout was covered by an algal crust. Erosion patches were formed in this crust between spring and early summer of 1987 (photo 3). A new algal crust, dominated by green alga, was established in August.

2.3.2 Large blowouts

In large blowouts algal crusts are mainly formed periodically because most new cruts are destroyed by wind erosion soon after their formation. Those algal crusts which remain intact for a longer time cover specific areas of the blowout. Fig. 2 relates to the period of the time during which algae are present at pin locations of blowout L2 and the proportional amount of green algae compared to cyanobacteria found there. The data indicate that algae are seldom found on the steep inner flank. On the other hand, algal crusts at pin 6.2, 5.3, 4.6 and 4.6 remain intact for a relatively long time. Two pins located outside the blowout (e.g. pin 4.4 and 4.5) indicate that "grey soils" surrounding the blowout are covered by algal crusts almost all the time. This is also the case for many south-exposed slopes outside the blowouts.

According to the species composition in blowout L2, green algae are generally associated with the centre and the inner flank areas orientated to the north whereas the cyanobacteria are more associated with other flank orientations. *Crinalium epipsammum* is the most important cyanobacteria which accompanied the increase of the green algae. The algal crusts outside the blowout contain comparatively more cyanobacteria.

In order to analyse the temporal development of the algal crust in large blowouts three types of algal crusts are distinguished, based on the development sequence of algae in small blowouts. For each observation period a figure is made which shows the surface level changes and the spatial extent of the crust type. Sequential interpretation of the algal colonization in blowouts reveals that the algal species composition partly depends on the time during which the algal crust remains intact and is able to proceed in the development sequence. Algal crust formation on the steep inner flank is frequently inhibited due to a high magnitude and frequency of sandtransport, for instance during early 1988 (fig. 3). Soil samples taken at pin 7.7, 7.8 and 7.9 contain generally no or only a few algae which belong mainly to the cyanobacteria, the pioneer colonizers. Cyanobacterial crusts formed here are eroded again before they can reach the green algae phase. Algal crusts are fully developed at pins which are less disturbed by sand dynamics, for instance at the sw to se side of the blowout. During the summer of 1986 the algal crust of blowout L2 was able to remain intact and could reach the mature stage during subsequent periods (fig. 4).

Analysis of the mechanism of algal colonization indicates that the species composition in blowouts is not primarily influenced by sand dynamics. Algal crusts on flanks located to the south seldom reach the mature green algal phase, even if there is hardly any sand transport during several months. This is also the case for algal crusts developed on "grey soils" surrounding the blowout as

algal crust development	frequency (%)	surface level change (mm)	s.d.	erosion (mm)
green algae → green algae	66	2.8	3.1	1.6
green algae → cyanobacteria	5	2.0	1.5	1.0
green algae → few algae	18	4.2	3.6	2.6
green algae → no algae	11	11.5	11.7	10.5
cyanobacteria → cyanobacteria	32	1.9	1.4	1.2
cyanobacteria → green algae	16	2.7	1.9	1.9
cyanobacteria → few algae	29	3.8	6.6	3.5
cyanobacteria → no algae	23	6.9	9.4	2.5
few algae → few algae	42	3.6	4.8	1.4
few algae → green algae	12	2.9	2.6	0.3
few algae → cyanobacteria	6	4.2	2.9	0.6
few algae → no algae	40	8.5	9.6	6.6

Tab. 1: *Alteration of three algal crust types during three years of study in blowout L2 and L3. The first column shows the development of the crust expressed proportionally to all alterations for that crust type. The other two columns indicate the mean surface level change and erosion rate which accompanied crust development.*

they tend to have comparatively greater amounts of cyanobacteria.

Tab. 1 shows the sand dynamics which accompanied the alteration of the algal crust types. The data refer to pins located inside the large blowouts L2 and L3. It is noted that the green algal crusts are still present at the next observation (after about one month) more frequently than with the other crust types. The frequency percentages reflect to some extent the successional pattern in which cyanobacteria become dominated by green algae. However, frequently cyanobacteria are not succeeded by green algae and only few algae or no algae at all are found at locations which were first covered by cyanobacterial crusts. The deflation rates of the three types of algal coverage do not significantly differ. The disappearance of all types of algal crusts is accompanied by relatively high sand dynamics. The values of the sand dynamics results from the resistance of sand against deflation due to algae and on the other hand from the influence of the sand dynamics on algal colonization.

The influence of algae on the erosion rate in large blowouts is indicated more properly by considering areas periodically covered by algae, for instance in blowout L3 (fig. 5). From 23/4-86 until January 1987 (period A) the algal surface contributed to a lowered deflation rate at the west part of this blowout. During the subsequent period B (6/1-87 to 26/3-89) no algae were present. The increased resistance of sand against deflation is clearly illustrated by comparing the surface level change on the west part with that on the east part for both periods. If only periods are taken into account when winds are blowing predominantly from the w and s, it appears that the mean ratio for period A (-0.42) is significantly lower than for period B (+0.60) when no algae are present (t = -5.87, sign. level = 0.001).

2.4 Discussion

Two kinds of colonization are distinguished: initial and secondary colonization. Initial colonization is by definition colonization of relatively large areas in a relatively short period of time which occurred for instance in blowout S1. This process is most important in the late summer or early autumn when wind velocities are generally low while heavy rains are responsible for deposition of sand with organic matter in the blowout. The organic matter stimulates initial colonization. Secondary colonization concerns the balance between algal colonization and sand dynamics as described for blowout L2. The most favourable conditions for a high algal growth rate are expected to occur during the period September to April or, more specifically, the period October to February in accordance with the course of the precipitation surplus throughout the year. Also the mean windspeed is commonly high at this time, so that the blowout will be subjected to colonization and deflation within a short time. These algal crusts generally cover relatively small areas in this period.

Small blowouts contain generally larger areas of algal crust. It is likely that the residual algal vegetation around erosion patches contributes to influxes of colonizing species. Especially during heavy summer rain storms overflowing water will contribute to influxes of colonizing species which can re-establish an algal crust. If, on the other hand, the algal vegetation is eroded, it reaches a state in which it can not respond any more to the stimulus of available precipitation. It is therefore assumed that the extent of an algal crust influences the rate of secondary blowout stabilization.

Deflation stops almost completely in small blowouts after algae appear. In large blowouts algae lower the erosion rate as well. However, a combination of biomass measurements and pinreadings in large blowouts indicate that some algal crusts still reveal deflation (tab. 1). An explanation for this might be that algal colonization occurred soon after erosion within the period that preceded soil collecting.

Although crusts with green algae remain intact for a longer period of time compared to cyanobacterial crusts, it appeared that when green algae were eroded the surface lowering was comparatively greater. This might be explained by the fact that initial colonization starts during periods of favourable soil water conditions and low windspeeds. If cyanobacteria are succeeded by green algae during the subsequent period, the influence of wind might increase.

The establishment of the algal crust appeared to be greatly influenced by sand dynamics but species composition in addition depends on other environmental factors. It is assumed that the succession at the centre of the blowout and on south exposed slopes outside the blowout is taking place relatively slowly due to unfavourable environmental conditions there. The results of a preliminary study concerning the spatial soil water content in blowouts indicate that soil conditions here are generally unfavourable compared to most of the blowout area. Therefore another aspect might even be more important and enable the cyanobacteria to adapt to the most adverse conditions existing on the slopes orientated to the south inside and outside the blowout. Contrary to *K. flaccidum* the two cyanobacteria *Tychonema* and *C. epipsammum* in study are able to restart photosynthesis after desiccation very quickly as soon as water is available (DE WINDER et al. 1989).

Various properties enable the cyanobacteria to act as better pioneer species than the green algae. At low light intensities most cyanobacteria have a higher net growth yield than green algae (VAN LIERE & MUR 1979). This allows them to grow in extremely low light intensities below the top sand layer because of their low maintenance energy requirements. Cyanobacteria might adapt its depth beneath the surface to the amount of sand deposition, as, except for *C. epipsammum*, all the isolated species are able to move phototactically and so it allows them to keep contact with the photic zone. Apparently this may enable the cyanobacteria to be pioneer colonizers in the moving sand.

References

BIRSE, E.M., LANDSBERG, S.Y. & GIMINGHAM, C.H. (1957): The effects of burial by sand on dune mosses. Trans. Br. Bryol. Soc. 3, 285–301.

BOUGHEY, (1957): Ecological studies of tropical coast-lines. I. The Gold Coast, West Africa. J. Ecology 45, 665–687.

CHAPMAN, V.J. (1964): Coastal Vegetation. MacMillan Co., New York, 245 pp.

DE WINDER, B., MATTHIJS, H.C.P. & MUR, L.R. (1989): The role of water retaining substrata on the photosynthetic response of three drought tolerant phototrophic micro-organisms isolated from a terrestrial habitat. Arch. Microbiology 152, 458–462.

DE WIT, R. & VAN GEMERDEN, H. (1988): Interactions between phototrophic bacteria in sediment ecosystems. Hydrobiological bulletin 22(2), 135–145.

FORSTER, S.M. (1979): Microbial aggregation of sand in an embryo dune system. Soil Biochem. 11, 537–543.

GRAEBNER, P. (1910): Pflanzenleben auf den Dünen. Dünenbuch. F. Enke Verlag, Stuttgart, 183–296.

GRANT, J. & GUST, G. (1987): Prediction of coastal sediment stability from photopigment content of mats of purple sulphur bacteria. Nature 330, 244–246.

HERBERT, R.A. (1985): Development of mass blooms of photosynthetic bacteria on sheltered beaches in Scapa Flow, Orkney Islands. Proc. Roy. Soc. of Edinburgh 87b, 15–25.

MORTON, J.K. (1957): Sand dune formation on a tropical shore. J. Ecology 45, 495–497.

NOEST, V. (1987): Stabilisatie van stuifkuilen. De rol van vegetatie en dynamiek van het substraat. Doctoraalverslag Fysisch Geografisch en Bodemkundig Lab., Universiteit van Amsterdam.

PLUIS, J.L.A. & DE WINDER, B. (1989): spatial patterns in algae colonization of dune blowouts. CATENA 16, 499–506.

PYE, K. (1980): Beach salcrete and eolian sand transport: evidence from North Queensland. J. Sed. Petrol. 50, 257–261.

REYNAUD, P.A. & ROGER, P.A. (1981): Variation saisonnières de la flore algale et de l'activite fixatrice d'azote dans un sol engorgé de bas de dune. Rev. Ecol. Bil. Sol. 18.

RIPPKA, R., DERUELLES, J., WATERBURY, J.B., HERDMAN, M. & STANIER, R.Y. (1979): Generic assignments, strain histories and properties of pure cultures of cyanobacteria. J. Gen. Microbiol. 111, 1–61.

STAL, L.J., VAN GEMERDEN, H. & KRUMBEIN, W.E. (1985): Structure and development of a benthic marine microbial mat. FEMS Microbiol. Ecol. 31, 111–125.

SVASEK, J.N. & TERWINDT, J.H. (1974): Measurements of sand transport by wind on a natural beach. Sedimentology 21, 311–322.

VAN DER PUTTEN, W.H., VAN DIJK, C. & TROELSTRA, S.R. (1988): Biotic soil factors affecting the growth and development of *Ammophila arenaria*. Oecologia 76, 313–320.

VAN DEN ANCKER, J.A.M., JUNGERIUS, P.D. & MUR, L.R. (1985): The role of algae in the stabilization of coastal dune blowouts. Earth surface processes and landforms 10, 189–192.

VAN DIEREN, J.W. (1934): Organogene Dünenbildung. M. Nijhoff, Den Haag.

VAN LIERE, L. & MUR, L.R. (1979): Growth kinetics of *Oscillatoria agardhii* in continuous culture, limited in its growth by the light energy supply. J. Gen. Microbiol. 115, 153–160.

WESTHOFF, V. (1947): The vegetation of dunes and salt marshes on the Dutch Islands of Terschelling and Vlieland and Texel.

ZENKOVICH, Z.P. (1967): Aeolian processes on sea coasts. In: Steers, J.A. (ed.), Processes of Coastal Development. Interscience, New York, 586–617.

Addresses of authors:
J.L.A. Pluis
Fysisch Geografisch en Bodemkundig Laboratorium
Dapperstraat 115
1093 BS, Amsterdam
B. de Winder
Laboratorium voor Microbiologie
Nieuwe Achtergracht 127
1018 WS, Amsterdam

EUROPEAN DUNES: CONSEQUENCES OF CLIMATE CHANGE AND SEALEVEL RISE

F. Van Der Meulen, Amsterdam

Summary

The effects of sealevel rise on coastal processes are direct and indirect. Direct effects include loss of land, re-enforcement of foredunes, sanddrift from foredunes inland, and influence of salt-spray more inland. The main indirect effect is lowering of the groundwater table.

Climate change affects the interaction of geomorphological and biological processes in the dune ecosystem. A simple scheme in which plants (vegetation cover) are regarded as agents for geomorphologic and/or soil development is presented. An assumed sealevel rise of about 1 m/100 year and a doubling of the atmospheric CO_2 concentration are used to discuss the effects on the interaction of geomorphological and biological processes. Two well-known European coastal dune areas with contrasting impact of climate change (the Coto Doñana National Park in southwestern Spain and the Slowinski National Park in northern Poland), are used as an example.

Most processes affected by climate change and sealevel rise are poorly understood. Directions for future research are presented. This includes inventory and monitoring studies. Ecologically indicative species (bio-indicators) and abiotic target variables are suitable monitoring objects.

1 Introduction

We know from geological records that the sea level in Europe has been rising constantly since the end of the last glacial period (JELGERSMA et al. 1970). Relative rise in sealevel for the southern North Sea basin is estimated at 45 m since 10,000 BP, at 5 m since 5,000 BP and at 0.5 m since 1000 BP (ABRAHAMSE et al. 1976). The present trend is 10–20 cm per century. Most likely, the rise will accelerate as a result of climatic changes due to the "green house effect". This process may continue for decades even if all carbon dioxide emissions were to be stopped now (DELFT HYDRAULICS 1988)! In Europe, as in most other parts of the world, coasts are eroding for more than a century (BIRD 1985). This is attributed to a combination of natural sealevel rise, isostatic lowering of the land and human intervention in coastal processes.

Coastal dunes are known for their valuable ecosystems with high conservation values. They also harbour several important functions for society such as the protection of coastal lowlands which

ISSN 0722-0723
ISBN 3-923381-23-9
©1990 by CATENA VERLAG,
D–3302 Cremlingen-Destedt, W. Germany
3-923381-23-9/90/5011851/US$ 2.00 + 0.25

1)	economic efects (property and production losses, cost of measures)
2)	public health (safety, diseases, availability of food supplies)
3)	environmental/ecological (loss of nature areas, disruption of ecological systems)
4)	social (unemployment, resettlement, loss of public utilities)
5)	administrative (legal problems, competence and jurisdiction, administrative boundaries)

Tab. 1: *Main impacts of sea level rise for nature and society.*

are amongst the most densely populated and industrialized parts of Europe.

The impacts of sealevel rise can be divided into five main categories (tab. 1). This paper concentrates on the environmental/ecological impacts. Socioeconomic impacts are not discussed.

The following aspects will be dealt with:

- what is presently known about climate change and sealevel rise?

- what are the effects of sealevel rise on the coastline?

- what are the consequences of a general change in climate for the natural dune landscape?

- which avenues of research are necessary to increase awareness and develop adequate management strategies?

2 Present knowledge about climate change and sealevel rise

Using the hierarchical rank model of BAKKER et al. (1981), fig. 1 indicates the relations between landscape components, beginning with climate. Our present knowledge about how the climate is going to change and how this will affect the dune landscape is insufficient and speculative. So, this paper will indicate possible effects in a hypothetical manner.

The various climate models disagree about the climatic changes to be expected. Lack of information on factors such as feedback influence of clouds, ice albedo and ocean heat storage is given as an explanation (JAGER 1986). Also there are no data on extremes and changes in extremes of climatic parameters. (The presence and duration of extremes could be factors much more decisive for the ecosystems than the mean values). By consequence scenarios of sealevel rise also differ (fig. 2). VAN HUIS (1989) discussed effects of climatic change on some European dune areas. He used a climate scenario developed by BACH (1986) for the European Workshop on interrelated Bioclimatic and Land Use Changes. This scenario is based on the GISS-GCM program, developed by HANSEN et al. (1984). The climate parameters temperature, precipitation and evaporation are incorporated in this model under different CO_2 situations. The program postulates a global warming up of 1.5–5.5 °C in the next

Climate Change and Sealevel Rise

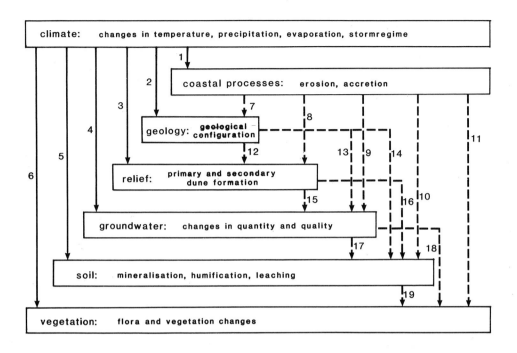

Fig. 1: *Main relations between landscape components, starting with climate. Feedback relations not shown. (after BAKKER et al. 1981).*

———> / - - - - > = direct / indirect relation;
1 = temperature, relative sealevel rise, change in storm regime;
2, 7 = speed of sealevel rise, erosion or accretion;
3, 8 = storm regime, phases of dune formation;
4 = change in amount and/or distribution of yearly evaporation;
5 = change in effective precipitation, temperature, sunshine, air humidity, soil formation;
6 = change in all climate factors;
9 = change in width of dunes causing change in groundwater level and influence of sea-spray;
10, 11 = drift-sand activity / wind erosion, change in seaspray;
12, 14 = lime content, mineral composition of sand, grain size;
13 = hydrogeology;
15 = microtopography groundwater table;
16, 17 = changes in groundwater regime, soil and vegetation;
18 = inundation or desiccation of valleys;
19 = succession, (de-)eutrophication, decalcification and acidification.

Global temp. rise	2–4°C
base trend	20 cm
thermal expansion	10–16 cm
glaciers	15–25 cm
Greenland ice	4–8 cm
Antarctica growth	-5–0 cm
	44–69 cm

Tab. 2: *Sea level rise caused by climate change due to a doubling of atmospheric CO_2 content (VAN HUIS 1989).*

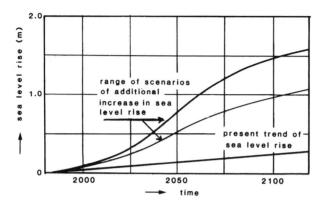

Fig. 2: *Scenarios of sealevel rise (DELFT HYDRAULICS 1988).*

century. The associated sealevel rise is approximately 100 cm, ranging rom 25–165 cm. As JELGERSMA (1979) indicated, the major factor contributing to this rise would be the thermal expansion of ocean water (tab. 2).

Statistical proof for sealevel rise resulting from an increased CO_2 content is not yet available (MIN. OF ROADS AND PUBLIC WORKS Oral Comm.).

Climate models do not take into account change in wind regime. Most likely, storm frequency and intensity will increase. But it is unknown as to what extent. An unfavourable change in wind regime causes bigger waves and extra erosion. In a recent study on coastal defence management in the Netherlands (MIN. OF ROADS AND PUBLIC WORKS 1989) different morphological scenarios are used to account for this effect in some way (see next chapter).

3 Consequences of sealevel rise for the Dutch coastline

The Netherlands are a low-lying country with densely populated areas immediately behind coastal dunes. Without the defence line of dunes and dykes almost 40% of the country would be flooded by the sea. Of the Dutch coast 255 km (about 80%) is protected by dunes. They harbour the following main landuse categories: nature conservation, public drinking water catchment, recreation, housing and industry. This paper focusses on the consequences of sealevel rise for nature.

Coastline changes have been monitored since about 1850 (relative position of the average low tide and high tide and the dune foot in relation to a reference line). Since 1963 coastline transects are measured. Results from several monitoring stations show a break in the trend of change around the year 1960. This break is probably due to great engineering works elsewhere along the coast.

Recently an integrated study was completed of the effects of sea level rise on changes of the coastline and its impacts for society (MIN. OF ROADS AND PUBLIC WORKS 1989). The period considered was one century: 1990–2090. On basis of this study, the Dutch government is preparing a new policy for coastal defence and management.

3.1 Morphological scenarios

To predict the effect of sealevel rise on the morphology of the coast a number of factors have to be taken into account. Basically two groups can be distinguished (VAN STRAATEN 1961, MIN. OF ROADS AND PUBLIC WORKS 1989):

a) geometrical conditions (such as the coast profile and civil technical works at the coast) and

b) hydraulic conditions (such as wind, currents, tides).

The exact influence of these conditions on the transport capacity of onshore and alongshore currents is not yet known. Tides and waves are master factors for sand transport at the coast. Wave incidence is an important factor for longshore transport, wave climate for transport in onshore and offshore directions.

Climatic change not only involves warming up and sealevel rise but will also affect storms and waves. To predict the effect for coastline changes, three morphological scenarios were used. They assume a sealevel rise of respectively 20 cm, 60 cm and 85 cm per century. To account for the uncertainty in storm frequency and wave regime these were divided in an unfavourable and a favourable scenario. The former expects both onshore and alongshore transport of sand to become 20% more unfavourable. This means that the onshore transport will decrease with 20% (leaving less sand to be transported onto the beach), while the alongshore transport increases with 20% (this means that more sand will leave a given coastal sector). A sealevel rise of 20 cm per 100 year and an unfavourable morphological scenario is comparable to 85 cm sealevel rise under favourable conditions.

The relation between coastal erosion/accretion and alongshore sediment transport is important also for other coasts. ROHRLICH & GOLDSMITH (1984) developed a sediment transport model of the eastern Mediterranean coast. They argue that no changes in the longshore transport of sediment along the Israeli coast are anticipated as a result of the Aswan Dam. The loss of sediment discharge into the sea due to this dam is more than adequately offset by erosion of the Nile Delta beaches.

The coastal dunes south of Valencia (Spain) are expected to suffer from erosion when the sealevel will rise, because there is no sediment supply along this section of the coast at the moment (SANJAUME & PARDO 1989).

3.2 Direct consequences for the natural landscape

Direct consequences of sealevel rise for the natural landscape are summarized below.

- Loss of land

  The Dutch study (MIN. OF ROADS AND PUBLIC WORKS 1989) showed that, if no action would be taken, considerable losses of the natural dune landscape will occur particularly in the Wadden Island Region in the north. These losses will be appreciable already at the start of the next century. They are due to coastline retreat and subsequent lowering of the phreatic level (fig. 3). The dune systems mainly affected are primary and secondary dunes with dune slacks and heath.

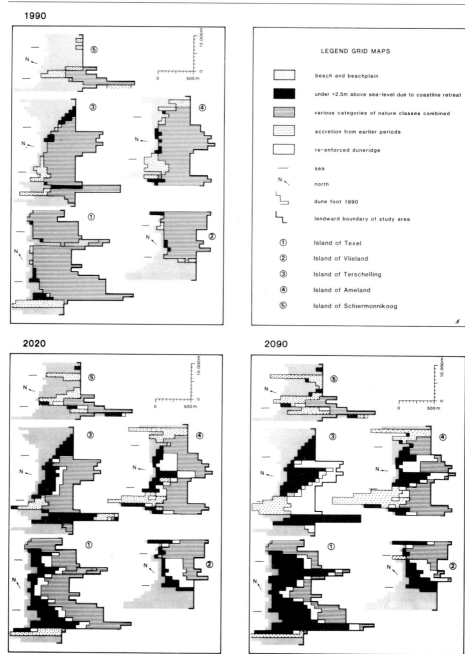

Fig. 3: *Losses of nature as a result of coastline retreat and lowering of groundwater level for five dutch Wadden Islands, for the years 1990, 2020, 2090 (after MIN. OF ROADS AND PUBLIC WORKS 1989).*

Data are stored in a Geographical Information System with grid size of 1000 m (alongshore) × 50 m (inland). North Sea is at north arrow (left in all figures). Note coastline retreat at Terschelling (3) and Texel (1); accretion at Ameland (4); Schiermonnikoog (5) and Vlieland (2) more or less stay the same but re-enforcements have to be made.

Loss of land not necessarily implies a loss of natural values. This will depend on the potentials for nature development in the new situation. Marine rejuvenation of the landscape may create a varied coastline with tidal inlets and marshes alternating with dry coastal ecosystems. Many dutch dune ecologists consider this a positive development after a long time of artificial stabilisation (cf. VAN DER MEULEN & VAN DER MAAREL 1989).

- Re-enforcement of foredune ridge

This is a deliberate action by the manager to ensure a safe defence against the sea. To re-enforce the foredune ridge, parts of the original landscape have to be buried by sand. This means extra loss of nature.

- Sanddrift from foredunes inland

The effect of this geomorphological process is difficult to quantify. The amount of sand transported inland is unknown as is the tolerance of plants and vegetations to slight amounts of sand burial. Some pioneer plants are known to be stimulated in their growth by accumulating sand, for example *Ammophila arenaria* (marram grass) on the white dunes and *Corynephorus cansescens* (grey hair grass) on the grey dunes. *Ammophila* can tolerate changes of 10–20 cm per year or more (HUISKENS 1977, VAN DER PUTTEN 1989); *Corynephorus* less, up to 5 cm per year. Some typical dune dwarfshrubs also occur in accumulation zones, notably *Ligustrum vulgare* (privet) and *Hippophaë rhamnoides* (sea buckthorn) (NOEST 1987). The author's own field observations show that these plants tolerate about 50 cm accumulation of sand per year.

- Salt spray

Airborne salts from the sea, notably chlorides, have a strong effect on dune vegetation and also on ionconcentrations in the groundwater. Studies on the island of Voorne between 1962/64 and 1979 showed a clear decline in halophytic plants of the foredunes after construction of extensive harbour facilities in front of the dunes. A lowering of airborne salt concentration may explain this (VEELENTURF 1982). The ecological importance of salt spray and the geomorphological consequences have not yet been quantified.

- Inundation

Inundation of land by sea water can only be assessed in its full extent when the topography of the dune is exactly known.

3.3 Indirect consequences for the natural landscape

The main indirect consequences of coastal erosion is a lowering of the groundwater level in the dune. The geohydrological equilibrium in the dune body will change; not directly but after some time. Height and depth of the fresh water body alter, as will the curves of the phreatic level and the fresh/brackish interface. The relation between coastline retreat and lowering of the phreatic level is given in fig. 4. Dune slack vegetations have very critical maximum and minimum heights of the phreatic level (VAN DER LAAN 1979). Changes of 10 cm, especially in the growing season, can be vital.

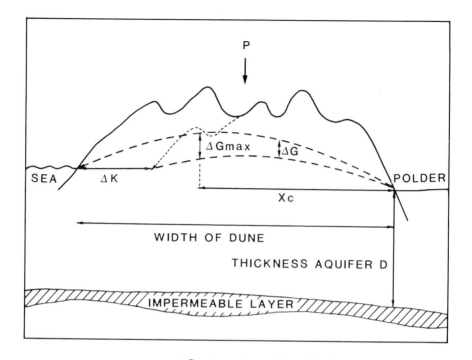

$$\Delta G = \frac{P * \Delta K * X}{2kD}$$

$$Xc = \frac{2kD * \Delta Gmax}{P * \Delta K}$$

ΔG = lowering phreatic level at position X
P = net input of precipitation
ΔK = coastline retreat
k = hydraulic conductivity of dune sand
D = thickness aquifer
X = distance from polder
$\Delta Gmax$ = maximum tolerable lowering of ΔG at distance Xc from polder

Fig. 4: *Relation between phreatic level and rate of coastline retreat (after MIN. OF ROADS AND PUBLIC WORKS 1989).*

Lowering of the phreatic level will also affect the geomorphological processes in the dunes. Excavating of dune valleys by the wind will go deeper when wet sand, which stops the process, is met at greater depth.

The effect of sealevel rise on the groundwater level is twofold. Erosion and narrowing of the dune strip causes a decline; on the other hand the rising sealevel causes a rise in groundwater level. To what degree the former is compensated by the latter effect depends on the properties of the dune area, the amount of erosion and the amount of sealevel rise.

Climate Change and Sealevel Rise

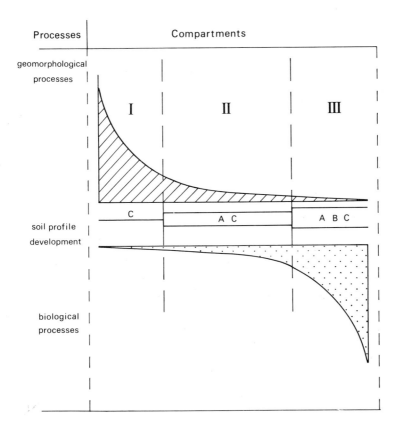

Fig. 5: *Schematic presentation of interaction of main processes of dune ecosystems (coastal processes and groundwater excluded) (JUNGERIUS & VAN DER MEULEN 1988).*

When the groundwater level in the upper aquifer of the dunes declines, the interface between fresh and brackish water in the deeper layers will rise at the same time (with a factor 20 to 40 as much). This is not of direct importance to the dune but it may lead to salt intrusion into the hinterland.

4 Consequences of general climate change for the dune ecosystem

4.1 Direct and indirect effects on the interaction of abiotic and biotic processes (development of relief, soils and vegetation)

A general climate change will affect the interaction between abiotic and biotic processes in the dune environment. The interplay between relief, soils and vegetation will change. Landscapes will develop either with more vegetation, mature soils and stable dunes, or with less vegetation, young soils and mobile dunes.

If we exclude coastal processes, basically two kinds of processes interact in the dune landscape: geomorphological and biological (mainly vegetation development and plant biomass production) (JUNGERIUS & VAN DER MEULEN 1988). No attempt is made here to quantify either of these processes. Fig. 5 merely shows the type and the relative importance of the processes. The interaction results in a particular configuration of landscape elements. They can be seen in the field in compartments. Interrelations between the compartments are not taken into account for the sake of simplicity.

Compartment I is dominated by geomorphological processes (wind erosion and overland flow). Consequently vegetation development is minor and soils are incipient or absent. In compartment III vegetation development is very pronounced. Organic matter production is high. Soil profiles are more developed (ABC profile). This affords resistance against erodibility: plants break the wind and humose soils erode less easily. Therefore geomorphical processes will be of minor importance in this compartment. The soil at any given place is an expression of the interaction between the two kinds of processes.

Climatic changes favouring vegetation development will impede geomorphological processes. A more "green" and stabilized dune landscape with "fossil" dune forms could be the result. Likewise, changes impeding vegetation development will favour the geomorphological processes. A more barren and dynamic landscape with active formation of new dune forms could be the result. In recent history both such situations occurred in Dutch dunes due to changes in land use (cf. VAN DER MEULEN & VAN DER MAAREL in press).

4.2 Coto Doñana National Park (Spain) and Slowinski National Park (Poland)

VAN HUIS (1989) discussed the possible impact of the climate changes on two well known European dune areas: the Coto Doñana in southern Spain and Slowinski National Park at the Baltic coast in Poland. They illustrate different kinds of impact in north and in south Europe (fig. 6).

Coto Doñana

The expected climate changes for the Coto Doñana in the south of Europe are as follows:

- the dry season will extend because of a mean monthly temperature rise of 4°C with no compensating increase in precipitation;

- the summer soil moisture decreases by 60%, leading to extreme drought;

- potential evaporation increases with 0.5mm/day during the winter months.

The simulated climate somewhat resembles the present-day climate of northern Libya, except that the drought period is shorter than in Libya and that average rainfall is much higher.

Ecological and geomorphological consequences of an increase in drought in combination with an expected sealevel rise of ca 100 cm per century are more coastal erosion and an increase in sanddrifts. A change to Mediterranean plant

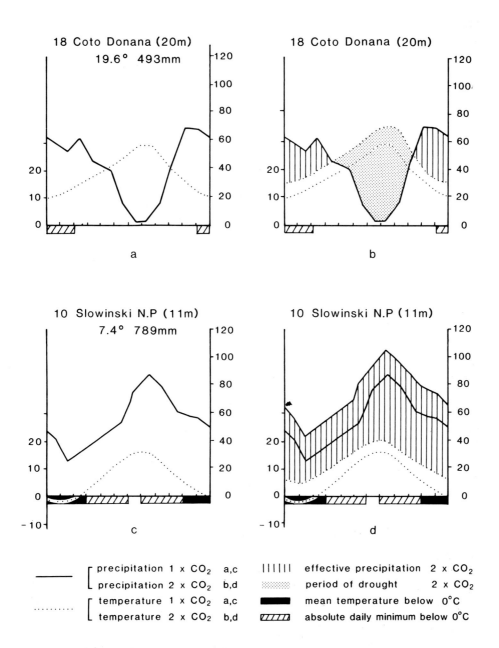

Fig. 6: *Ecological climate diagrams for Coto Doñana and Slowinski dune areas. Actual situation (a,c) with simulated 2X CO_2 concentration as overlay (b, d) (VAN HUIS 1989).*

species composition is possible, because Mediterranean plants appear to have a higher tolerance for water stress than oceanic species (MERINO 1986). The earlier start of drought (in April) will affect plant growth and establishment and consequently the vegetation succession. A higher evapotranspiration, especially in evergreen mediterranean Pines, will lead to lowering of the phreatic level and water deficiency in the soil. Forests will suffer and decline. A general decline in vegetation cover, a diminishing soil fertility and an impeded sand fixation will lead to desertification of the dune area.

More drought and a decrease of vegetation cover could stimulate geomorphological processes to become so intense that new aeolian dune forms develop: the situation would move to the left in fig. 5.

The wetlands (Marismas) behind the dunes may gradually become intruded with oceanic water and change into estuaries. This will affect the avifauna. The Coto Doñana is an important foraging resource for birds migrating from northern to southern Europe and Africa.

Slowinski National Park

For Slowinski National Park at the Baltic coast in the north of Europe it is expected that the present-day subcontinental climate will become more atlantic. Winter temperatures will increase by 5°C. Regular frosts will rarely occur any more in December and April. Rainfall will increase as will the number of humid summer months. Evaporation will increase though this will be partly compensated by an increase in precipitation. Comparable climates are currently found around the North Sea, but these have lower precipitation in summer.

Ecological and geomorphological consequences of an increase in humidity together with an expected sealevel rise of approximately 1 m (the continental uplift of the Fennoscandian Shield will not occur in the south-east part of the Baltic sea (JELGERSMA 1979)) are as follows. The vegetation cover will extend over barren parts of the dunes. Wind erosion will decrease. *Corynephorus canescens*, preferring a non-calcareous substratum, could spread over the grey dunes and over the then immobile white dunes. Plant species composition will change in favour of deciduous trees associated with higher rainfall, higher temperatures and more acid soil conditions (birch, oak, alder). Deciduous woodlands will increase and could possibly bring the famous mobile dunes to a standstill. A shortening of the freezing period will affect the water temperature of the dune lakes and the aquatic fauna.

The more humid conditions and the increase in vegetation cover could lead to a situation where geomorphological processes have largely come to a stop: the situation moves towards the right in fig. 5.

5 Future research

Two groups of research can be recognized: inventory and monitoring.

5.1 Inventory

Corre (LICC 1989) proposes the preparation of maps (scale about 1:5000) depicting the vulnerability of plant and geomorphological indicators for the short, medium and the long term (years, decade and century respectively). The legend of such maps should be based on a classification of the vulnerability of these indicators to the effects of climate change such as sealevel rise and wind erosion.

A survey of the European dune areas, their extent, physiographic characteristic and ecological quality, is urgently needed. Special reference should be given to those areas threatened by sealevel rise either because of serious erosion and changes in groundwater regime or because of inundation since they lie below 1–2 m above average sealevel. Beach plains, primary dune forms, wetlands and tidal marshes belong to these environs. Such a survey is being initiated by the European Union for Dune conservation and Coastal management (EUDC 1989). Progress of this important activity is hampered by a lack of funds.

It is also advocated to store and handle data in a Geographic Information System (LICC 1989). Digitizing of information is necessary. Units of 50×50 m appear to be appropriate to discern in the field because landscape ecological patterns in the dunes manifest themselves already at this scale. Larger units generalize too much. They also render it impossible to locate exactly where a specific ecological relationship is affected (for example groundwater level change and vegetation response).

Information rapidly gets out of date. Especially vegetation maps do. Maps at scale of 1:5000 up to 1:10 000 need to be updated every 5–10 years. Traditional mapping implies much work and problems of generalisation. It pays to use scanners and special software to transform information from aerial photos directly into a GIS. This will facilitate sequential surveys.

5.2 Monitoring

RITCHIE (LICC 1989) suggests the following for the monitoring of environmental change:

- careful selection of the target variable that provides the best index of change;

- careful selection of control sites; statistically valid mechanisms that can distinguish between real changes and normal cyclic perturbations such as occur in all natural systems.

Target variable:

RITCHIE prefers the selection of a specific indicator for climate change in the vegetation or the geomorphology. As geomorphological target variables he suggests sand movement in relationship to wind velocity and wind direction; total sediment availability and changes in relief amplitude and landform orientation. Sequential aerial photography and other image interpretation appear to be promising, together with actual measurements for ground truth.

There is a lack of causal and quantitative information on the interaction of the biotic and the abiotic environment. Ecological interaction or response models should be developed. Master abiotic variables to be monitored for this purpose should ideally comprise groundwater and soil moisture, atmospheric deposition and soil chemistry. Rainfall, air temperature and radiation seem usefull climate variables to monitor.

Site selection:

According to RITCHIE the best sites are those where variation of other system variables can be held to a minimum or be controlled. There should be no change in past, existing and future landuse inside nor outside the system. For example, introduction of new plants and animals should not occur; drainage other than naturally controlled hydrology should not change. Outside the area no building of sea walls, groynes or other defence structures should occur in order not to alter sediment transport at the edge of the system.

The study of historical records may give an indication of activity, geographical area or dune type which is likely to be affected by changes in climate or sealevel.

When monitoring plant or animal species one should select those species which are most indicative to specific abiotic conditions ("bio-indicators") and which are of nature conservation value.

The distinction between changes that can reasonably be ascribed to climate change and natural changes that are cyclic and normal systems perturbations is another aspect to be taken into account when selecting sites and objects for monitoring.

Acknowledgements

I am grateful for the comments on drafts of this paper and for useful discussions with C. Louisse (Tidal Waters Division, Dutch Min. of Roads and Public Works), J.V. Witter (Univ. Amsterdam) and W. Ritchie (Univ. Aberdeen).

References

ABRAHAMSE, J., JOENJE, J. & VAN LEEUWEN-SEELT, N. (1976): Waddenzee. Ver. tot Behoud van Natuurmon. & Ver. tot Behoud van de Waddenzee, Harlingen. 368 pp.

BACH, W. (1986): GCM-derived climatic scenarios of increased atmospheric CO_2 as a basis for impact studies. In: Parry, M.L., Carter, T.L. & Konijn, N.T. (eds.), Assessment of climate impacts on Agriculture. Vol. **1**: High latitude regions. Reidel, Dordrecht.

BAKKER, T.W.M., KLIJN, J. & VAN ZADEL-HOFF, F.W. (1979): Duinen en duinvalleien, een landschapsecologische studie van het Nederlandse duingebied. Pudoc, Wageningen, 201 pp.

BAKKER, T.W.M., KLIJN, J. & VAN ZADEL-HOFF, F.J. (1981): Nederlandse kustduinen. Landschapsecologie, Pudoc, Wageningen, 144 pp.

BIRD, E.C.F. (1985): Coastline changes a global review. Wiley & Sons, New York, 219 pp.

DELFT HYDRAULICS (1988): Sea level rise, a global issue. General information leaflet.

EUDC (1989): Program Plans 1989–1992. European Union for Dune conservation and Coastal management. Leiden, The Netherlands.

HANSEN, J., LACIS, A., RIND, D., RUSSELL, G., STONE, P., RUEDY, J. & LERNER, J. (1984): Climate sensitivity: analysis of feedback mechanisms. In: Hansen, J. & Takahasi, T. (eds.), Climate processes and climate sensitivity. Maurice Ewing Serie **5**. Amer. Geoph. Un., 130–163.

HUISKENS, A.H.L. (1977): The population dynamics of *Ammophila arenaria*. Ph.D. Thesis, Univ. of Wales.

JAGER, J. (1986): Climatic change: Floating new evidence in the CO_2 Debate. Environment **28** (7), 6–9 and 38–41.

JELGERSMA, S. (1979): Sealevel changes in the North Sea basin. In: Oele, et al., The Quaternary history of the North Sea. Acta Univ. Uppsala Symp. Univ. Upps. Annum quingentesimum Celebrantis 2, 233–248.

JELGERSMA, S., DE JONG, J., ZAGWIJN, W.H. & VAN REGTEREN ALTENA, J.F. (1970): The coastal dunes of the western Netherlands: geology, vegetational history and archaeology. Meded. Rijks Geol. Dienst. Haarlem. N.S.P. 93–167.

JUNGERIUS, P.D. & VAN DER MEULEN, F. (1988): Erosion processes in a dune landscape along the Dutch coast. CATENA **15**, 217–218.

LICC (1989): Landscape ecological impacts of climatic change in coastal dunes of Europe. Working Report for LICC (Landscape Ecological Impact of Climatic Change) conference by F. van der Meulen & P.D. Jungerius. Lunteren, The Netherlands, 171 pp.

MERINO, J. (1986): Desertification in Spain: importance of man and climate as triggers. EC-MAB Symposium on desertification in Europe.

MINISTRY OF ROADS AND PUBLIC WORKS (1989): Ministerie van Verkeer & Waterstaat. Discussienota kustverdediging na 1990.

The Hague. 20 Technical Reports. References from report nrs 5, 6, 8.

NOEST, V. (1987): Stabilisatie van stuifkuilen. De rol van vegetatie en dynamiek van het substraat. MSc. Report Univ. of Amsterdam / Dune Water Works. The Hague.

ROHRLICH, V. & GOLDSMITH, V. (1984): Sediment transport along the southeastern mediteranean: a geological perspective. Geo-Marine Letters Vol. 4, 99–103.

SANJAUME, E. & PARDO, J. (1989): Sealevel rise impact on the precarious dunes of Devesa del Saler beach. Valencia, Spain. Invited paper to LICC conference Lunteren, The Netherlands. 13 pp.

VAN DER LAAN, D. (1979): Spatial and temporal variation in the vegetation of dune slacks in relation to the groundwater regime. Vegetatio 39, 43–51.

VAN DER MEULEN, F. & VAN DER MAAREL, E. (1989): Coastal defence alternatives and nature development perspectives. In: F. Van Der Meulen, P.D. Jungerius & J. Visser (eds.), Perspectives in coastal dune management. 183–195. SPB Acad Publ. The Hague.

VAN DER MEULEN, F. & VAN DER MAAREL, E. (in press): Coastal dunes of the southern and central Netherlands. In: E. Van Der Maarel (ed.), Dry coastal Ecosystems of the World. Elsevier, Amsterdam.

VAN DER PUTTEN, W.H. (1989): Establishment, growth and degeneration of *Ammophila arenaria* in coastal sand dunes. Thesis Agric. Univ. Wageningen.

VAN HUIS, J. (1989): European dunes, climate and climatic change, with case studies of the Coto Donana (Spain) and the Slowinski (Poland) National Parks. In: F. Van Der Meulen, P.D. Jungerius & J. Visser (eds.), Perspectives in coastal dune management. 313–326. SPB Acad. Publ. The Hague.

VAN STRAATEN, L.M.J.U. (1961): Directional effects of winds, waves and currents along the Dutch North Sea coast. Geologie en Mijnbouw 40e Jaargang, 333–346 and 363–391.

VEELENTURF, P. (1982): Saltspray. Rep. Dept. Physical Geogr. Univ. Utrecht. 109 pp.

Address of author:
F. Van Der Meulen
Landscape and Environmental Research Group
University of Amsterdam
Dapperstraat 115
1093 BS Amsterdam
The Netherlands

INDISPENSABLE for your scientific advancement!

INDEX: CATENA VOL.1-15, CATENA SUPPLEMENT 1-13, 1973 - 1988

336 pages/ DM 16.-/US$ 10.-

Dear scientist,

CATENA is proud to offer you a comprehensive INDEX, 1973 -1988 at an extremely reasonable price of DM 16.-/US$10.- (336 pages).

The obvious success of CATENA over the last fifteen years is proof of the enormous demand for the subject matter by both authors and readers. It will continue to promote up-to-date process-oriented research with relevance to contemporary landscapes without neglecting the historical dimension. The subtitle of **CATENA**, **"GEOECOLOGY AND LANDSCAPE EVOLUTION"** expresses this duality of contemporary processes and inherited structures, which should in reality be a unified approach - both aspects are relevant to the landscape.

Please, order your personal copy immediately.

Yours sincerely,

Margot Rohdenburg

Managing Editor CATENA

ORDER FORM

☐ Please send me:.........copy/ies: INDEX CATENA AND CATENA SUPPLEMENTS 1973 - 1988 at the price of DM 16.-/ US$10.- each

Prepayment is required

☐ check enclosed ☐ payment is enclosed by UNESCO coupons ☐ please charge my credit card

☐ MasterCard/Eurocard/Access ☐ Visa ☐ Diners ☐ American Express

Card No: .... Expiration date: ....

Date/Signature ....

Name ....

Address ....

Please send your orders to: CATENA VERLAG, Brockenblick 8, D-3302 Cremlingen-Destedt, West Germany

Orders from USA/Canada to: CATENA VERLAG, P.O.Box 368, Lawrence, KS 66044, USA

Epilogue

Although coastal dunes in Europe are widespread, they form a narrow and vulnerable zone. Apart from direct destruction and degradation by man's activities there are a number of natural threats to their existence, such as sea level rise and climatic change. Scientifically dunes are invaluable and deserve protection against these hazards. They harbour a rich combination of relatively undisturbed geological, geomorphological, hydrological, pedological and ecological features. In addition the dunes represent an extensive archeological and historical archive.

In this volume abiotic processes dealing with land forms, groundwater and soils were approached from various vantage points. Knowledge of these processes is a prerequisite for understanding the dune landscape. This book shows the 'state of the art' in various countries and points to promising avenues of research. From these contributions a picture arises of further abiotic research that is needed on a European scale.

1. A serious handicap for **all disciplines** is the fact that a comprehensive picture of abiotic features of the European coastal dunes is not yet available. This limits the degree to which research findings can be transferred from site to site. Priority should therefore be given to European research programmes that aim at a complete inventory of the geomorphological forms, hydrological systems, ecosystems and soils, using systematic classifications embracing useful national and regional approaches. The inventories presently carried out within the framework of the Conservation Programme of the European Union for Dune Conservation and Coastal Management (EUDC) and the CORINE Programme are steps in the right direction, but the level of detail of the collected abiotic information is attuned to dune managers and decision makers, and is therefore not sufficiently refined to serve the purpose of scientific investigations. For further fruitful research we should aim at a higher level of resolution.

   Another impediment for all disciplines is the lack of information on processes. This can only be overcome by systematic monitoring. A fairly detailed proposal for establishing a number of stations along the European coast for monitoring the effect of sealevel rise and climatic change was recently put forward by the European Conference on the Landscape-ecological Impact of Climatic Change (LICC) held in The Netherlands at the end of last year. The proposed monitoring programmes require the cooperation of national dune management organizations and the coordination of the data set at the European level. Such initiatives merit encouragement.

2. In **geomorphology** there is a clear need for a better understanding of the interactions between **geomorphological processes** and the biotic components of the landscape. The role of algae in the stabilization of blowouts discussed in this volume is a case in point.

   Age and origin of coastal dune systems throughout Europe form intriguing subjects for further study. Contributions in this volume as well as publications elsewhere deserve further analysis of datings from a correlative point of view. There should also be attention for the diverging hypotheses of dune formation in different regions. These hypotheses should be related to other sources of geological and climatological history and to our insight of man's influence in the past.

3. **Hydrological systems** in coastal dunes can differ greatly due to difference in climate, underlying geology and influences of the hinterland. In this volume, several approaches applicable to small areas are presented. As fresh-water reservoirs in dunes are of great importance for man and nature, such studies should be extended.

4. **Soil studies** of coastal dunes are scarce. A soil classification focussed on dune ecosystems and incorporating the specific combination of pedological, geomorphological and hydrological dynamics has yet to be designed. This is essential for mapping. In addition there is a great need for process studies. The effects on soils and vegetation of acid deposition and atmospheric supply of nutrients are largely unknown and ask for field study, laboratory experiments and modelling. Various aspects dealt with in this volume are far from exhausted such as the micromorphology of dune soil organic matter and the waterrepellency exhibited by dune sand and its relevance for plant growth and erosion.

What form should future research take? There is no blueprint for answering this question. No doubt the gowing political awareness that ecological problems have to be tackled on a European level will lead to further research funding by EC organizations. More than anything else this will enable us to work together in coordinated programmes such as STEP and EPOCH. Some of the other paths presently availabe are:

- to arrange special conferences such as those organized by the EUDC and Eurocast;

- to organise workshops dealing with particular subjects or problems;

- to publish special volumes, such as the present CATENA SUPPLEMENT or the Proceedings of the European Dune Symposium held in Leiden in 1987.

We are convinced that we express the wish of all of the authors of this volume if we encourage our European colleagues in the 'CATENA' sciences to explore these and other ways of realizing our goals. The unique landscape of the European coastal dunes deserves our continuing effort to further research that transgresses our individual national boundaries.

Theo Bakker
Peter Jungerius
Jan Klijn

Appendix

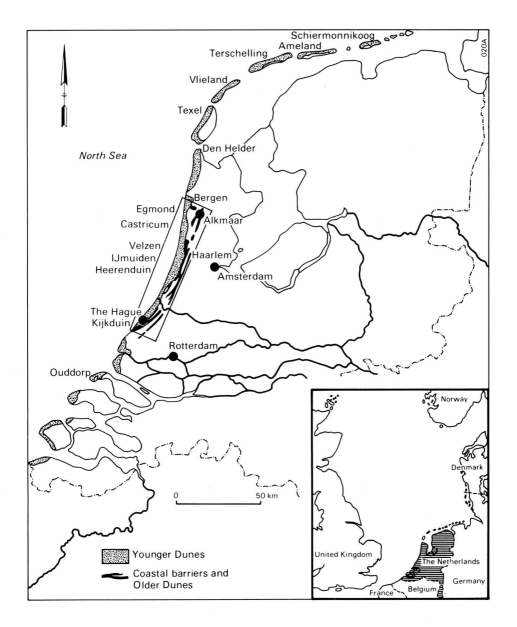

Coastal dunes in The Netherlands; names refer to the contributions of i.a. KLIJN (2), JUNGERIUS and DEKKER. The rectangle indicates the area where dune deposits have been dated by ^{14}C (see KLIJN).

NEW PUBLICATION 1990

SPECIAL INTRODUCTORY OFFER 30% OFF

valid until November 30, 1990

R.B.Bryan (Editor)

SOIL EROSION
— Experiments and Models —

CATENA SUPPLEMENT 17

hardcover/about 230 pages/numerous figures, photos and tables

ISSN 0936-2568/ISBN 3-923381-22-0

list price: DM 139.-/US $ 75.-/standing order price CATENA SUPPLEMENTS: DM 97,30/US$ 52.50

ORDER FORM

☐ Please send me at the special introductory rate of DM 97,30/US$ 52.50 .... copies of CATENA SUPPLEMENT 17.

☐ I want to subscribe to CATENA SUPPLEMENTS (30% reduction on the list price) starting with no.

Name ....

Address ....

Date ....

Signature: ....

Please charge my credit card: ☐ MasterCard/Eurocard/Access ☐ Visa ☐ Diners ☐ American Express

Card No.: .... Expiration date: ....

Please, send your orders to:

CATENA VERLAG, Brockenblick 8, D-3302 Cremlingen-Destedt, West Germany, tel.05306-1530, fax 05306-1560

USA/Canada:**CATENA VERLAG**, Attn. Kristi Koloukis, P.O.Box 368, Lawrence, KS 66044, USA, Tel. (913) 843-1234, fax (913) 843-1244

CATENA paperback

Joerg Richter
THE SOIL AS A REACTOR
Modelling Processes in the Soil

If we are to solve the pressing economic and ecological problems in agriculture, horticulture and forestry, and also with "waste" land and industrial emmissions, we must understand the processes that are going on in the soil. Ideally, we should be able to treat these processes quantitatively, using the same methods the civil engineer needs to get the optimum yield out of his plant. However, it seems very questionable, whether we would use our soils properly by trying to obtain the highest profit through maximum yield. It is vital to remember that soils are vulnerable or even destructible although or even because our western industrialized agriculture produces much more food on a smaller area than some ten years ago.

This book is primarily oriented on methodology. Starting with the phenomena of the different components of the soils, it describes their physical parameter functions and the mathematical models for transport and transformation processes in the soil. To treat the processes operationally, simple simulation models for practical applications are included in each chapter.

After dealing in the principal sections of each chapter with heat conduction and the soil regimes of material components like gases, water and ions, simple models of the behaviour of nutrients, herbicides and heavy metals in the soil are presented. These show how modelling may help to solve problems of environmental protection. In the concluding chapter, the problem of modelling salt transport in heterogeneous soils is discussed.

The book is intended for all scientists and students who are interested in applied soil science, especially in using soils effectively and carefully for growing plants: applied pedologists, land reclamation and improvement specialists, ecologists and environmentalists, agriculturalists, horticulturists, foresters, biologists (especially microbiologists), landscape planers and all kinds of geoscientists.

Prof.Dr. Joerg Richter
Institute of Soil Science
University of Hannover, FRG

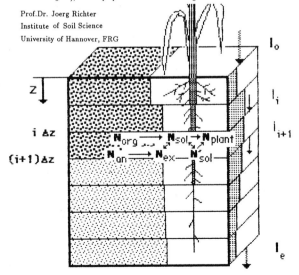

ISBN 3-923381-09-3   Price: DM 38,50 / US $ 24.—

Heinrich Rohdenburg

LANDSCAPE ECOLOGY — GEOMORPHOLOGY

CATENA paperback

1989/about 220 pages/DM 44.-/US$ 28.-

ISBN 3-923381-15-8

"The outline shows that geomorphology could obviously occupy an important place in this geoecology. "**Morphogenesis and ecology!**" should replace "morphogenesis or ecology?".

For this to happen, **processes must be placed at the centre of research**, not only the present-day observable and measurable processes but also past processes and their environmental conditions which can be derived by substrate analysis. These past conditions can be reconstructed from an investigation of the substrate.

Analysis of processes and of the substrate ought to be given greater weight in geoecology, although not at the expense of analysis of the land surface. This means that the explanatory value of the latter is, in fact, enhanced by the inclusion of the weak linkage between the land surface and the substrate.

Relationships with **soil science** and with **hydrology** should be intensified in this connection because of the relevance of both subjects to geomorphology."

ORDER FORM

☐ Please send me .... copies of: Heinrich Rohdenburg, **LANDSCAPE ECOLOGY — GEOMORPHOLOGY**, CATENA paperback. 1989 at the rate of DM 44.-/ US $ 28.-

☐ **English version** ☐ **Original German text**

Name ....

Address ....

Date ....

Signature: ....

Please charge my credit card: ☐ MasterCard/Eurocard/Access ☐ Visa ☐ Diners ☐ American Express

Card No.: .... Expiration date: ....

Please, send your orders to:

CATENA VERLAG, Brockenblick 8, D-3302 Cremlingen-Destedt, West Germany, tel.05306-1530, fax 05306-1560

USA/Canada:**CATENA VERLAG**, Attn. Denize Johnson, P.O.Box 368, Lawrence, KS 66044, USA, Tel. (913) 843-1234, fax (913) 843-1274

CATENA

AN INTERDISCIPLINARY JOURNAL OF

SOIL SCIENCE
HYDROLOGY - GEOMORPHOLOGY

FOCUSING ON

GEOECOLOGY AND LANDSCAPE EVOLUTION

founded by H. Rohdenburg

A Cooperating Journal of the International Society of Soil Science.

ISSS-AISS-IBG

CATENA publishes original contributions in the fields of

GEOECOLOGY, the geoscientific-hydro-climatological subset of process-oriented studies of the present ecosystem,
- the total environment of landscapes and sites
- the flux of energy and matter (water, solutes, suspended matter, bed load) with special regard to space-time variability
- the changes in the present ecosystem, including the earth's surface

and

LANDSCAPE EVOLUTION, the genesis of the present ecosystem, in particular the genesis of its structure concerning soils, sediment, relief, their spatial organization and analysis in terms of paleo-processes;
- soils: surface, relief and fossil soils, their spatial organization pertaining to relief development,
- sediment with relevance to landscape evolution, the paleo-hydrologic environment with respect to surface runoff, competence, and capacity for transport of bed material and suspended matter, infiltration, groundwater and channel flow,
- the earth's surface, relief elements and their spatial-hierarchical organization in relation to soils and sediment
- the paleoclimatological properties of the sequence of paleo-environments.

CATENA publishes multidisciplinary studies as well as monodisciplinary papers that are of interest to other disciplines and are of relevance to landscape studies.

CATENA VERLAG

ISSN 0341-8162